Review Criteria for Successful Treatment of Hydrolysate at the Blue Grass Chemical Agent Destruction Pilot Plant

Committee on Review Criteria for Successful Treatment of Hydrolysate at the
Pueblo and Blue Grass Chemical Agent Destruction Pilot Plants

Board on Army Science and Technology

Division on Engineering and Physical Sciences

The National Academies of
SCIENCES · ENGINEERING · MEDICINE

THE NATIONAL ACADEMIES PRESS
Washington, DC
www.nap.edu

THE NATIONAL ACADEMIES PRESS • 500 Fifth Street, NW • Washington, DC 20001

This activity was supported by Contract No. W911NF-14-1-0280 with the U.S. Department of Defense. Any opinions, findings, conclusions, or recommendations expressed in this publication are those of the author(s) and do not necessarily reflect the views of the organizations or agencies that provided support for the project.

International Standard Book Number-13: 978-0-309-37640-2
International Standard Book Number-10: 0-309-37640-8

Additional copies of this report are available for sale from the National Academies Press, 500 Fifth Street, NW, Keck 360, Washington, DC 20001; (800) 624-6242 or (202) 334-3313; http://www.nap.edu.

Printed in the United States of America

Suggested citation: National Academies of Sciences, Engineering, and Medicine. 2015. *Review Criteria for Successful Treatment of Hydrolysate at the Blue Grass Chemical Agent Destruction Pilot Plant*. Washington, DC: The National Academies Press.

The National Academies of

SCIENCES · ENGINEERING · MEDICINE

The **National Academy of Sciences** was established in 1863 by an Act of Congress, signed by President Lincoln, as a private, nongovernmental institution to advise the nation on issues related to science and technology. Members are elected by their peers for outstanding contributions to research. Dr. Ralph J. Cicerone is president.

The **National Academy of Engineering** was established in 1964 under the charter of the National Academy of Sciences to bring the practices of engineering to advising the nation. Members are elected by their peers for extraordinary contributions to engineering. Dr. C. D. Mote, Jr., is president.

The **National Academy of Medicine** (formerly the Institute of Medicine) was established in 1970 under the charter of the National Academy of Sciences to advise the nation on medical and health issues. Members are elected by their peers for distinguished contributions to medicine and health. Dr. Victor J. Dzau is president.

The three Academies work together as the **National Academies of Sciences, Engineering, and Medicine** to provide independent, objective analysis and advice to the nation and conduct other activities to solve complex problems and inform public policy decisions. The Academies also encourage education and research, recognize outstanding contributions to knowledge, and increase public understanding in matters of science, engineering, and medicine.

Learn more about the National Academies of Sciences, Engineering, and Medicine at **www.national-academies.org**.

COMMITTEE ON REVIEW CRITERIA FOR SUCCESSFUL TREATMENT OF HYDROLYSATE AT THE PUEBLO AND BLUE GRASS CHEMICAL AGENT DESTRUCTION PILOT PLANTS

TODD A. KIMMELL, Argonne National Laboratory, Washington D.C. Office, *Chair*
EDWARD J. BOUWER, Johns Hopkins University, Baltimore, Maryland
JUDITH A. BRADBURY, Pacific Northwest National Laboratory (retired), Knoxville, Tennessee
REBECCA A. HAFFENDEN, Argonne National Laboratory, Santa Fe, New Mexico
KIMBERLY L. JONES, Howard University, Washington, D.C.
MURRAY GLENN LORD, Dow Chemical Company, Freeport, Texas
DOUGLAS M. MEDVILLE, The MITRE Corporation (retired), Highlands Ranch, Colorado
TRISHA H. MILLER, Sandia National Laboratories, Shoreview, Minnesota
ROBERT B. PUYEAR, Union Carbide (retired), Chesterfield, Missouri
WILLIAM R. RHYNE, ABS Consulting, Inc. (retired), Kingston, Tennessee
PHILLIP E. SAVAGE, Penn State University, University Park, Pennsylvania
PHILIP C. SINGER, NAE, University of North Carolina, Chapel Hill (retired)
SETH TULER, Worcester Polytechnic Institute, Massachusetts

Staff

JAMES C. MYSKA, Study Director
NIA D. JOHNSON, Senior Research Associate
DEANNA P. SPARGER, Program Administrative Coordinator

Preface

The Blue Grass Chemical Agent Destruction Pilot Plant (BGCAPP) is being constructed under the direction of the Program Executive Officer for Assembled Chemical Weapons Alternatives (PEO ACWA). BGCAPP is scheduled to begin its operations to destroy chemical munitions in 2018. Following munitions access and hydrolysis of nerve agents and energetics, BGCAPP will use a first-of-a-kind (FOAK) technology known as supercritical water oxidation (SCWO) to treat the hydrolysate and a water recovery system (WRS) to recover water for reuse as quench water in the SCWO reactors. Except for scheduled maintenance, the SCWO and WRS systems are intended to operate continuously for over 3 years.

In an attempt to address potential issues with the systems that will be used to treat hydrolysate, BGCAPP will conduct preoperational testing. This testing will take place concurrent with plant systemization and will address process and equipment problems that were identified during and after FOAK testing. These measures should help to address anticipated problems as BGCAPP begins the actual munitions destruction process. The SCWO and WRS systems should be able to be operated successfully.

Yet there remain a number of factors that could cause the SCWO and WRS to underperform. For example, the agent and energetic hydrolysates to be generated at BGCAPP constitute a very complex matrix. As another example, the actual SCWO system that will be used at BGCAPP has never processed actual agent or energetic hydrolysates. Further, the SCWO and the WRS have never been operated together. These and other factors could lead to the SCWO and WRS underperforming and possibly delay the destruction of the chemical munitions stored at the Blue Grass Army Depot. These munitions have now been stored for over five decades. Many of them have leaked and, although the chemical munitions are stored in a protective manner, continued storage of these munitions represents an ongoing risk to the local community. These munitions must be destroyed safely and efficiently, and it is therefore important to ensure that problems with the SCWO and WRS do not delay this process. All the more so since there will be long lead times associated with implementing any alternatives to processing hydrolysates through the SCWO and WRS. The Army needs a backup plan for treating the BGCAPP hydrolysate and needs to be in a position to implement it expeditiously if necessary.

I would like to thank the PEO ACWA and BGCAPP staff and systems contractors who provided input to the committee's deliberations and accommodated its numerous inquiries. I also want to thank the Kentucky Department for Environmental Protection and the local Citizens' Advisory Commission and the Chemical Destruction Community Advisory Board for offering their perspectives on the issues. I must also thank the staff of the National Academies of Sciences, Engineering, and Medicine for their tireless and outstanding support, especially Jim Myska, Deanna Sparger, and Nia Johnson. Lastly, I thank the committee members for putting up with my onerous demands, challenging schedule, and my dry and only sometimes witty sense of humor.

Todd A. Kimmell, *Chair*
Committee on Review Criteria for Successful Treatment of Hydrolysate at the Pueblo and Blue Grass Chemical Agent Destruction Pilot Plants

Acknowledgment of Reviewers

This report has been reviewed in draft form by individuals chosen for their diverse perspectives and technical expertise, in accordance with procedures approved by the Report Review Committee. The purpose of this independent review is to provide candid and critical comments that will assist the institution in making its published report as sound as possible and to ensure that the report meets institutional standards for objectivity, evidence, and responsiveness to the study charge. The review comments and draft manuscript remain confidential to protect the integrity of the deliberative process. We wish to thank the following individuals for their review of this report:

Martin A. Abraham, Youngstown State University,
Cheryl Burke, The Dow Chemical Company,
Edward L. Cussler, Jr. (NAE), University of Minnesota, Minneapolis,
Julius Rebek, Jr. (NAS), The Scripps Research Institute,
Bruce E. Rittmann (NAE), Arizona State University
Leonard M. Siegel, Center for Public Environmental Oversight,
Vernon L. Snoeyink (NAE), University of Illinois,
William J. Walsh, Pepper Hamilton LLP.

Although the reviewers listed above have provided many constructive comments and suggestions, they were not asked to endorse the conclusions or recommendations nor did they see the final draft of the report before its release. The review of this report was overseen by Hyla S. Napadensky (NAE), who was responsible for making certain that an independent examination of this report was carried out in accordance with institutional procedures and that all review comments were carefully considered. Responsibility for the final content of this report rests entirely with the authoring committee and the institution.

Contents

APPENDIXES

Tables, Figures, and Boxes

TABLES

FIGURES

BOXES

Acronyms and Abbreviations

ACWA	Assembled Chemical Weapons Alternatives
AFS	aluminum filtration system
APG	Aberdeen Proving Ground
APS	aluminum precipitation system
atm	atmosphere (unit of pressure)
BGAD	Blue Grass Army Depot
BGCAPP	Blue Grass Chemical Agent Destruction Pilot Plant
CAC	Citizens' Advisory Commission
CATT	Citizens' Advisory Technical Team
CDCAB	Chemical Destruction Community Advisory Board
CWC	Chemical Weapons Convention
CWWG	Chemical Weapons Working Group
DoT	U.S. Department of Transportation
DSCM	dry standard cubic meter
EBH	energetics batch hydrolyzer
EIS	environmental impact statement
EPA	U.S. Environmental Protection Agency
FEIS	final environmental impact statement
FOAK	first-of-a-kind
GB	nerve agent (sarin)
gpm	gallons per minute
HPGLS	high-pressure gas–liquid separator
HSA	hydrolysate storage area
HVAC	heating, ventilation, and air conditioning
KAR	Kentucky Administrative Regulations
KDEP	Kentucky Department for Environmental Protection
KEF	Kentucky Environmental Foundation
LDR	Land Disposal Restriction
LPGLS	low-pressure gas–liquid separator
mil	one one-thousandth of an inch
NECD	Newport Chemical Depot
NECDF	Newport Chemical Agent Disposal Facility
NEPA	National Environmental Policy Act
NRC	National Research Council
OPCW	Organisation for the Prohibition of Chemical Weapons
OWS	oil–water separator
PCAPP	Pueblo Chemical Agent Destruction Pilot Plant
PCB	polychlorinated biphenyl
PCD	Pueblo Chemical Depot
PEO	Program Executive Office
ppm	parts per million
psig	pounds per square inch gauge
QTRA	quantitative transportation risk analysis
RCRA	Resource Conservation and Recovery Act
RD&D	Research, Development, and Demonstration
REC	record of environmental consideration
RMA	Rocky Mountain Arsenal
RO	reverse osmosis
SCWO	supercritical water oxidation
SDC	static detonation chamber
SDS	spent decontamination solution
SFT	shipping and firing tube
TDS	total dissolved solids
TOC	total organic carbon
TSCA	Toxic Substances Control Act

TSDF	treatment, storage, and disposal facility
VX	nerve agent
WRS	water recovery system
wt %	percentage by weight, or weight percent

Summary

One of the last sites with a stockpile of chemical munitions is the Blue Grass Army Depot (BGAD), located near Richmond, Kentucky. The stockpile at BGAD consists of rockets and projectiles that contain the nerve agents GB (sarin) and VX and the blister agent mustard. The rockets also contain energetics, including a burster explosive and propellant. Under the direction of the Program Executive Office (PEO) for Assembled Chemical Weapons Alternatives (ACWA), the Blue Grass Chemical Agent Destruction Pilot Plant (BGCAPP) is in the final stages of construction and will destroy the munitions containing GB and VX.[1,2]

Destruction of the chemical weapons is mandated by the Chemical Weapons Convention (CWC).[3] BGCAPP operations will be overseen by the Kentucky Department for Environmental Protection (KDEP), which administers the environmental permits under which BGCAPP must operate. Additional BGCAPP stakeholders include the public, represented by the Kentucky Citizens' Advisory Commission (CAC) and the Chemical Destruction Community Advisory Board (CDCAB).

Caustic hydrolysis will be used at BGCAPP to destroy the agents and energetics, resulting in a secondary waste stream known as hydrolysate. A first-of-a-kind (FOAK) technology, supercritical water oxidation (SCWO), will be used to treat the hydrolysates. SCWO will mineralize organic materials in the hydrolysates by reacting them in water above its critical temperature and pressure (T_c = 705°F, P_c = 218 atm (3,200 psi)). SCWO effluent will then be sent to a water recovery system (WRS) that will use reverse osmosis (RO) to separate salts from water. The end products of the SCWO and WRS processes are water and brine. The water is intended for reuse as SCWO quench water, reducing the amount of water needed from onsite sources. The brine will be sent offsite for disposal. Treatment processes are detailed in Chapter 2.

Considering the FOAK nature of SCWO and the complexity of the hydrolysate, ACWA has been concerned that SCWO and WRS may not function as designed. ACWA commissioned two earlier National Research Council studies of the BGCAPP systems, one for the WRS, completed in 2012, and another for SCWO, completed in 2013.[4] The authoring committees made a number of recommendations, but there were no overarching concerns that the processes would not work, assuming their recommendations were adequately addressed. ACWA has commissioned this study by a committee of the National Academies of Sciences, Engineering, and Medicine[5] to further examine the possibility of delay or failure of the SCWO or WRS and to examine possible alternatives to onsite treatment. The Statement of Task for this study is provided in Chapter 1.

As discussed in Chapters 6 and 7 of this report, BGCAPP will conduct preoperational testing concurrent with facility systemization intended to identify and resolve problems with equipment and operating procedures. This testing is expected to increase the likelihood of success. However, underperformance or failure of the SCWO and/or WRS could result in BGCAPP not meeting its overall performance criteria, as set forth in Chapter 6. The remainder of this Summary discusses the content of the report, directing the reader to the pertinent chapters for more detailed discussion, and presents selected findings and recommendations that highlight what the committee believes are the key points of the report. Also,

[1] BGCAPP is called a pilot plant because some of the processes used for destroying the agent have never before been used in this application, or used in combination with each other.

[2] The mustard stockpile at BGAD will be disposed of onsite using an explosive destruction technology; it will not be processed through BGCAPP.

[3] Convention on the Prohibition of the Development, Production, Stockpiling and Use of Chemical Weapons and on Their Destruction. The treaty entered into force in 1997.

[4] *Letter Report: The Blue Grass Chemical Agent Destruction Pilot Plant's Water Recovery System*, 2012, and *Assessment of Supercritical Water Oxidation System Testing for the Blue Grass Chemical Agent Destruction Pilot Plant*, 2013 (NRC, 2012; 2013).

[5] Effective July 1, 2015, this institution, formerly sometimes referred to as the National Research Council, is called the National Academies of Sciences, Engineering, and Medicine.

as explained in Chapter 1, the committee did not develop original success criteria for the SCWO and WRS as called for in the statement of task. While new criteria could be developed, these should be based on preoperational testing and systemization, which for BGCAPP has yet to be initiated. In the course of its work, this committee received briefings from PEO ACWA, BGCAPP staff, public stakeholders, and KDEP during its January 2015 meeting in Kentucky, which took place at two locations near BGCAPP. The committee has since reviewed extensive documentation on the BGCAPP processes and engaged in significant back and forth with BGCAPP staff and public stakeholders.

Finding 1-1. It is expected that extensive preoperational testing of the SCWO and the WRS will be performed concurrently with systemization to help reduce the uncertainty in expected technological performance. However, considering the first-of-a-kind nature of these technologies and the need for them to properly function in tandem, there is a possibility that technological issues with these systems may prevent BGCAPP from meeting all of its overall performance criteria.

Chapter 7 provides a detailed discussion of technical factors that may lead to underperformance of the SCWO or WRS. If problems occur, the hydrolysis process may have to be interrupted. In the event that hydrolysis is halted, and as hydrolysate storage nears capacity, destruction of the stockpile at BGAD may need to be halted unless there is an alternative means for treating the hydrolysate. The nerve agent munitions at BGAD have been stored for over 50 years. Delays in the destruction process will protract the risk to the community associated with continued storage. The committee believes that destruction of the BGAD stockpile must continue to eliminate the risk. Hence, the committee believes it is necessary to establish a backup plan.

Recommendation 1-1. Considering that the SCWO and the WRS may not perform satisfactorily and that this underperformance could result in delays in the destruction of chemical agent at BGCAPP, increasing the primary risk to the community associated with continued munitions storage, BGCAPP should establish a backup plan as an alternative to the onsite hydrolysate treatment processes.

Should SCWO or WRS problems develop, substantial storage capacity exists at BGCAPP, allowing time to improve performance while agent destruction continues. Hence, the likely scenario if issues develop would be that BGCAPP would have time to resolve any SCWO or WRS problems as neutralization continues. Only if hydrolysate storage nears capacity, or if problems appear unsolvable, would it become necessary to consider the backup plan. One alternative that would likely be more readily implemented than others would be to ship the hydrolysate offsite for treatment and disposal.

Further, if the WRS underperforms or fails, neutralization and SCWO operations could continue as long as BGCAPP can obtain permits to withdraw additional water from local sources for SCWO reactor quenching. BGCAPP would then need to seek alternative means of disposal of the SCWO effluent. Alternatively, BGCAPP could forgo SCWO and send the hydrolysate offsite instead. Continuing SCWO operations would destroy compounds regulated by the CWC, obviating further oversight by the Organisation for the Prohibition of Chemical Weapons (OPCW) if untreated hydrolysate were to be shipped offsite. This option, however, would entail a significant increase in offsite waste shipments, as discussed in Chapter 5. The benefits of further CWC oversight will need to be considered against the costs of increased waste shipments.

The choices that will be faced with possible underperformance of SCWO or the WRS are highly complex. All alternatives will entail interconnected considerations about technological capabilities and limitations, in conjunction with PEO ACWA commitments, contractual requirements, public stakeholder opinion, regulatory permitting requirements, schedule delays, and cost.

STAKEHOLDER CONCERNS

The concerns of the community are a key consideration in any offsite decision process. The community around BGCAPP is represented primarily by the CAC and the CDCAB. The CAC comprises nine members appointed by the state governor. An independent CAC subcommittee, the CDCAB, provides for broader stakeholder involvement from local organizations including medical, emergency management, university, and school representatives.

The committee held its first meeting in Richmond, Kentucky, in January 2015 and invited CAC/CDCAB members to attend. The committee also held a separate public meeting during the January 2015 sessions. The CAC chair and one co-chair of the CDCAB attended the 2-day open meetings and also attended the public meeting.[6] As part of the discussions and presentations at the meeting, the CAC/CDCAB provided written and verbal statements of their views on the conduct of the study, including criteria that they believe should guide decisions regarding potential offsite shipment. Several conference calls were also conducted, both with CAC/CDCAB representatives and with PEO ACWA and BGCAPP public involvement officials. As a result of these activities, the committee learned that despite initial conflicts, the CAC/CDCAB, PEO ACWA, and BGCAPP have built a solid working relationship. Should operational issues arise at BGCAPP this working relationship will provide a strong basis for continued and meaningful public involvement.

[6] The position of the second co-chair was in transition at the time of the committee's site visit.

The CAC/CDCAB recognizes that the weapons at BGAD must be destroyed as soon as and as safely as possible. They view onsite treatment with hydrolysis followed by the SCWO and WRS technologies as a commitment made to the community. As documented in previous reports the CAC/CDCAB opposed offsite hydrolysate shipment in the past (NRC, 2008; Noblis, 2008; NRC, 2012). They nevertheless recognize that there are bound to be issues with a pilot plant such as BGCAPP and that failure of key processes could force an examination of alternatives. The CAC/CDCAB maintain that offsite hydrolysate shipment would only be acceptable if it were the sole alternative, and should be temporary until BGCAPP systems are back online. Most important for the CAC/ CDCAB, however, is that they continue to be involved in the decision process.

Recommendation 3-1. In collaboration with the CAC/CDCAB, the Program Executive Office for Assembled Chemical Weapons Alternatives should institutionalize a transparent consultation process that builds on the existing foundation and working group structure to ensure meaningful stakeholder input into analyses, evaluations, and decision criteria related to potential offsite shipment of hydrolysate and that provides opportunities for engaging with communities that would receive hydrolysate.

REGULATORY ISSUES

Chapter 4 discusses regulatory requirements for potential offsite shipment of hydrolysate or SCWO effluent. Applicable regulations include the Resource Conservation and Recovery Act (RCRA), as administered by KDEP and the U.S. Environmental Protection Agency, and the National Environmental Policy Act (NEPA). There are also state and local requirements regarding the transport of hazardous materials, and also water withdrawal permits that need to be considered. In addition, the requirements of the CWC, as administered by the OPCW, need to be considered.

RCRA permitting requirements provide the greatest challenge for BGCAPP. The challenge is made greater by the unique situation in Kentucky, where legislation establishes hazardous waste listings for the chemical agents. The legislative language pertaining to chemical agent would classify waste derivatives such as hydrolysate and SCWO effluent as acutely toxic wastes, the same as the original agent. This acute toxicity designation imposes stringent requirements for management of these wastes that are burdensome and unnecessary. The CAC/CDCAB, in coordination with BGCAPP, are presently pursuing legislative changes that would alleviate this situation.

Another complication in Kentucky stems from the manner in which the BGCAPP RCRA permit is structured. BGCAPP is permitted under a RCRA Research Development and Demonstration (RD&D) permit for destruction of GB, but is intended to switch to a conventional RCRA Part B permit for destruction of the VX. Should SCWO be abandoned during the RD&D phase, the RD&D permit would no longer be applicable and BGCAPP would have to switch to a Part B permit. More important, should hydrolysate or even SCWO effluent need to be shipped offsite, under either the RD&D or a Part B permit, a year or more would likely be needed to process the required permit modifications.

Recommendation 4-4. As a backup plan, BGCAPP should revise its RCRA Part B permit application currently being prepared for the disposal of VX munitions to allow for the possibility of offsite transport of VX agent, GB agent, and energetics hydrolysates, as well as spent decontamination solution and SCWO effluent should the SCWO or WRS process be shown to be irreparable.

Recommendation 4-5. As a backup plan, BGCAPP should consult with KDEP concerning whether the RCRA RD&D permit could be modified to allow the temporary offsite transport of GB hydrolysate (i.e., until the SCWO can be brought back on line) or for the temporary or permanent offsite transport of SCWO effluent should the WRS process be shown to be irreparable.

Further, NEPA documentation may be required to support transporting hydrolysate or SCWO effluent offsite. The option to transport these materials for treatment and disposal was not directly considered in the BGCAPP NEPA analyses that have been conducted. The NEPA process could also be lengthy and may also take many months if it becomes necessary to produce additional documentation.

The key point is that regulatory delays not impact the destruction of agent at BGCAPP. Accordingly, the committee believes that RCRA permit modifications and NEPA documentation that support the potential for offsite shipment as a backup plan need to be prepared expeditiously and discussed with all stakeholders.

TRANSPORTATION ASSESSMENT

One of the concerns regarding potential offsite transportation is the risk of a transportation crash or release incident. Chapter 5 summarizes previous offsite shipments from chemical demilitarization facilities. That summary demonstrates that hydrolysate and similar fluids have been shipped offsite from a number of locations without incident many times in the past.

That is not to say that a transportation crash or incident could not occur, however. But it is important to understand that the hazard posed by hydrolysate is not from the presence of agent within the hydrolysate: During the hydrolysis process, the agent is destroyed, and destruction is verified prior to SCWO treatment. Although the presence of agent degradation products, including CWC schedule 1 and 2 chemicals, is a concern, the primary hazard of a release of hydrolysate during transportation would come from its causticity. BGCAPP hydrolysate would be considered a Department of Transportation Class 8 (corrosive) material.[7] The hazards due to hydrolysate exposure are modest compared to exposure to materials such as concentrated sodium hydroxide, a typical Class 8 material.

There is also a possibility, as indicated in Chapters 4, 6, and 7, that SCWO effluent could be sent offsite. If SCWO effluent is sent offsite for further treatment or disposal, the hazard posed by this material itself would be minimal. The effluent is a non-toxic brine. The main risk of SCWO effluent shipment would come from a substantial increase in the number of offsite shipments and the associated increase in the likelihood of a crash.

While the committee believes that the transport of hydrolysate or SCWO effluent would be low-risk, it is desirable that PEO ACWA perform a quantitative transportation risk assessment, including a quantitative assessment of the human health consequences if there is a release of hydrolysate or SCWO effluent. It is also desirable that PEO ACWA prepare a prototypical emergency response plan. These documents will help facilitate discussions with the public and regulators about possible offsite shipment. As with regulatory documentation, the transportation assessment and emergency response plan should be prepared as a backup plan and be ready to go if it is determined that offsite transport of hydrolysate or SCWO effluent is needed.

CRITERIA FOR SUCCESSFUL HYDROLYSATE TREATMENT AND DECISION FRAMEWORK

Chapter 6 presents treatment criteria for agent and energetic hydrolysates and provides a framework for decision making. Two categories of criteria are defined in Chapter 6:

- *Performance requirements.* Conditions that must be met under regulatory permits and under CWC treaty obligations, and
- *Performance goals.* Primarily oriented toward process performance and schedule and should be met to enable satisfactory system performance.

Performance requirements are crucial in achieving successful BGCAPP operation. If these requirements cannot be met, and if the time it takes to destroy munitions increases, risk reduction goals will not be achieved in a timely manner, and the offsite hydrolysate option becomes a more attractive option.

Finding 6-1. The primary criteria for successful treatment of hydrolysate involve meeting regulatory and Chemical Weapons Convention requirements and meeting process performance and schedule goals for hydrolysate treatment.

Performance goals consist mostly of quantitative expectations for SCWO and WRS performance. These goals are based on the results of past testing, modeling, and analysis by BGCAPP and its contractors. The committee anticipates that these goals may be modified during preoperational testing. A failure to meet some or many of the goals, while it may impact process performance, schedule, and costs, would not necessarily result in offsite shipment. However, if modifications made during systemization to achieve these goals do not result in improved performance, then consideration of offsite transport becomes more likely.

Recommendation 6-1. The ability to meet the initial performance goals established by BGCAPP for SCWO and the WRS should be verified as a result of testing during systemization.

Chapter 6 provides an example of graded evaluation to evaluate potential for success of SCWO and the WRS as the project proceeds toward completion. This is shown in Table S-1 (also Table 6-1).

TABLE S-1 Graded Success Scale for Use in Evaluating Overall Operation and Individual Treatment Processes (SCWO and WRS)

Grade	Definition
0	Success is practically certain (very low probability of SCWO or WRS failure): Operations are proceeding as expected. No BGCAPP actions needed.
1	High likelihood of success (low probability of SCWO or WRS failure): Actions should be taken by BGCAPP to prepare ahead of time for implementation of contingencies in the event of failures. For example, BGCAPP might begin to prepare permit modifications and planning documents.
2	Success is uncertain (moderate probability of SCWO or WRS failure): Actions should be taken to prepare for implementation of contingency operations. For example, BGCAPP might begin processing environmental documentation (permit modifications) and finalizing contingency plans and begin to initiate changes in infrastructure to permit off-site shipment.
3	Success is unlikely with current operations (high probability of failure of the SCWO or the WRS): Actions are taken to accelerate the implementation of contingency operations and stakeholders are consulted. For example, construction of needed facilities such as new piping and loading docks is completed as quickly as possible, environmental approvals are expedited, if not already obtained, and contracts for shipment offsite and disposal at a permitted treatment, storage, and disposal facility are signed.

[7] Class 8 hazmat is defined in 49 CFR 173.136 as a liquid or solid that causes (1) full thickness destruction of human skin within a specified period of time or (2) a specified corrosion rate of steel or aluminum.

Once hydrolysate processing begins, adjustments to SCWO and WRS operating procedures, process chemistry, and equipment will be made as needed. If the SCWO and the WRS fail to satisfy regulatory requirements and performance goals despite these adjustments, offsite hydrolysate shipment becomes more likely.

UNDERPERFORMANCE AND FAILURE RISKS, SYSTEMIZATION, AND CONTINGENCY OPTIONS

Chapter 7 discusses the possible risks of underperformance or failure of SCWO and WRS and focuses on decisions leading to possible changes in plant operations. The decision framework, the performance criteria, and the graded scale for success introduced in Chapter 6 are used in the discussion of these risks. The possibility of multicomponent failure is also discussed in Chapter 7.

Underperformance and Failure Risks, Systemization, and Contingency Options for the Supercritical Water Oxidation System

SCWO can irreversibly break down organic compounds using only oxygen, water, and supplementary fuel. Based on research and development of SCWO systems, including FOAK testing on simulated hydrolysates, BGCAPP scientists and engineers have identified a number of challenges that may be encountered in the operation of the SCWO system at BGCAPP and are taking appropriate corrective actions to address these challenges (BPBG, 2013).

Finding 7-1. BGCAPP scientists and engineers have identified a number of challenges that may be encountered in the operation of the SCWO system at BGCAPP and are taking appropriate corrective actions to address these challenges.

Nevertheless, the SCWO system has a number of components that all have to operate in unison. First, agent hydrolysate and energetics hydrolysate (after aluminum removal) are mixed along with specific amounts of various additives in the blend tank. In addition to requiring specific blend ratios of energetics hydrolysate to agent hydrolysate in the SCWO feed system, salt management is critical. Successful SCWO operation during agent processing will require close monitoring of feed composition with each feed campaign to ensure the target additive concentrations are reached prior to feeding the mixtures to the SCWO reactors.

Finding 7-4. The SCWO system is complex and has a number of components, all of which have to operate in unison for hydrolysate to be effectively destroyed in a timely manner.

Technical factors that may affect SCWO performance will be addressed during preoperational testing activities carried out during systemization. A Blue Grass SCWO Working Group was formed in April 2014 to address issues related to SCWO performance. The SCWO Working Group is addressing the gaps in knowledge, experience, and performance of the SCWO process, providing recommendations for closing these gaps, and producing a plan for implementing these recommendations. With this approach, the likelihood of reliable SCWO operations should be greatly enhanced.

Recommendation 7-3. The SCWO Working Group plan, as described in the December 17, 2014, Systemization Planning Report, and recommendations for correcting potential gaps in the October 27, 2014, Working Group report should be aggressively implemented. Furthermore, the SCWO Working Group should continue to provide support to all risk mitigation activities involving SCWO operations at BGCAPP.

While FOAK testing established reactor downtimes and maintenance cycles, these tests were only conducted on simulated hydrolysate streams and were conducted over relatively short periods of time compared to the expected 3-year destruction schedule at BGCAPP. There are contingency options if the SCWO does not meet performance requirements and goals, including materiel and procedural options. These are summarized in Table 7-1. If performance requirements and goals cannot be met consistently and satisfactorily, it may become necessary to consider shipping some or all of the hydrolysate or SCWO effluent offsite for disposal elsewhere.

Underperformance and Failure Risks, Systemization, and Contingency Options for the Water Recovery System

The WRS consists of three SCWO effluent storage tanks, a pretreatment system to remove suspended solids, three RO units (two in operation, one in reserve), two storage tanks for RO permeate, and two storage tanks for RO reject. The WRS influent will comprise two combined streams: (1) SCWO effluent and (2) cooling tower and steam blowdown water. These streams will be high in total dissolved solids, making RO an attractive process for treating the water so that it can be reused as quench water for the SCWO reactors. Although RO is an established treatment technology for high total dissolved solids waters, the SCWO effluent is a unique feed for RO and poses some treatment challenges. A prior National Research Council committee (2012) expressed the following concerns associated with the WRS system:

- Possible overloading of the pretreatment multimedia filters with particles from the SCWO process, including titanium dioxide, iron oxides, and calcium and aluminum phosphate precipitates;
- Potential for RO fouling and scaling due to inadequate pretreatment by the coagulation and direct filtration processes; and
- Durability of the materials of construction.

The committee believes that concerns expressed in the previous letter report have not been completely addressed in the current WRS design. The committee also believes that uncertainties associated with solids loading and coagulant requirements need to be addressed. It would be very useful for WRS performance to be assessed during pre-operational testing using realistic SCWO effluent simulants. These tests would ideally also establish operational conditions for effective filtration. If the influent solids concentration is too high, or if the actual loading threshold is far less than the design loading threshold, the filters could require frequent backwashing, increasing the demand for an alternative source of clean backwash water. The absence of a clarifier ahead of the filters increases the likelihood of overloading the filters with suspended solids.

Preoperational testing would also ideally test the effectiveness of inorganic coagulants for treating SCWO effluent and determine the exact conditions for effective coagulation. Further, it is uncertain how TiO_2 solids emanating from corrosion of the SCWO liner will respond to coagulation. Another potential problem is the volume and quality of water to backwash the filters. The SCWO effluent, the proposed source of backwash water, is expected to have a high suspended solids concentration, making this water source less than optimal for backwash water. A better option for backwash water for the multimedia filters would be the filter effluent itself.

Finding 7-13. Since many WRS process details are unknown, including the amount of solids in the SCWO effluent, the amount of solids that settle in the SCWO effluent storage tanks, and coagulant requirements and effectiveness, successful operation of the current WRS direct filtration multimedia pretreatment system is uncertain. Therefore, successful operation of the RO units is uncertain.

Recommendation 7-8. Well-planned preoperational testing should be performed with actual SCWO effluent, or a realistic simulant, to establish operating conditions for effective pretreatment and to determine if the WRS system, especially the multimedia direct filtration system, will perform as expected. In particular, preoperational testing should determine the solids loading and corresponding coagulant requirements for effective pretreatment. As noted in Chapter 6, serious consideration should be given to forming a WRS working group analogous to the SCWO Working Group.

In addition, the successful operation of the SCWO process for destruction of the agent hydrolysates is tightly linked to the production of quench water from the WRS units. If the WRS were unable to provide adequate quench water, another source of quench water for the SCWO system would be needed. There are contingency options if the WRS does not meet performance requirements and goals. These are summarized in Table 7-2. The challenge of complete WRS failure would be the need to seek alternative treatment or disposal of the SCWO effluent, which could involve an onsite process but, more likely, would entail offsite treatment or disposal.

Offsite Shipment as a Contingency Option

Chapter 7 discusses a number of potential modifications to SCWO and the WRS, weighing the causes and impacts of failure against the onsite contingency options. However, there may be scenarios in which SCWO system underperformance is so severe, compounded, or chronic that onsite mitigation actions are no longer sufficient. In this case, offsite shipment of hydrolysate would need to be considered. On the other hand, if the SCWO system performs adequately but the WRS underperforms or fails, there may be work arounds that enable BGCAPP to continue the destruction of hydrolysates using SCWO. In this case, BGCAPP would have the option to continue SCWO operations and send SCWO effluent offsite or halt both SCWO and WRS operations and send hydrolysate offsite. Continuing SCWO operations would meet CWC requirements and obviate the need for further OPCW oversight of hydrolysate sent offsite. Per Chapter 5, however, sending SCWO effluent offsite would substantially increase the number of offsite waste shipments. The resource requirements for increased offsite shipment would need to be carefully evaluated against the alleviation of further CWC oversight associated with continuing SCWO operations.

Should the decision be made to ship hydrolysate or SCWO effluent offsite, additional infrastructure would be needed to efficiently and effectively transfer the material for shipment. Infrastructure additions would likely be minimal but might include additional piping, leak containment, agent monitors, waste loading areas, truck loading docks, new rail/roadways onsite, new signage, and extra traffic controls at BGCAPP. As discussed in Chapter 5, shipping the hydrolysates offsite would actually produce fewer truckloads of hazardous waste material requiring disposal than the current plan.

Finding 7-16. A decision to ship hydrolysate offsite could have serious impacts on stakeholders, BGCAPP operations, regulatory compliance, and obligatory requirements under the Chemical Weapons Convention. There might be additional negative impacts if BGCAPP is not prepared ahead of time for a possible transition to offsite shipment, if and when such a decision is made.

The agent and energetics hydrolysates each have unique chemical compositions. In the current operating plan, the energetics hydrolysate is treated to remove aluminum and then blended with nerve agent hydrolysate prior to SCWO treatment. If offsite shipment is implemented, the committee believes it would not be necessary to remove aluminum from the energetics hydrolysate prior to offsite shipment, as long as the receiving treatment, storage, and disposal facility is able to accept the hydrolysate as is. It would also be just as acceptable to ship the hydrolysates separately as it would be to ship them after blending.

The committee also deliberated at length on whether the decision to ship offsite should be permanent, or if there are scenarios in which offsite shipment could be used temporarily or in parallel while the SCWO system and WRS operate at reduced availability. Implementing offsite transport of hydrolysate under any circumstance (temporary, parallel, or permanent) would affect plant, paper (e.g. regulatory permit modification), and people, as discussed in Chapter 7. Physical changes to the plant, changes in permit documentation and standard operating procedures, the conduct of transportation risk assessments, and staffing changes would need to be considered, just to name a few examples. Any effort to implement offsite transport would be considerable. Likewise, the effort to shift back to onsite treatment would also be substantial.

Offsite shipment in parallel with reduced onsite hydrolysate treatment might alleviate some of the transition burdens, but the scenarios in which this option is practical are limited. Also, operating SCWO at reduced capacity while also shipping some hydrolysate offsite would increase the system management burdens, in addition to those efforts required to repair the underperforming component(s). The committee acknowledges that at this time it is not possible to predict the exact circumstances of SCWO or WRS underperformance or failure once BGCAPP enters into operation and that the evaluation of whether to ship offsite permanently, temporarily, or in a parallel manner is more appropriately made by decision makers and stakeholders when the specific circumstances are known.

Throughout the report, the committee recommends that BGCAPP take actions, such as filing the necessary permit modifications and installing shipping infrastructure, necessary to prepare for the last-resort decision to ship hydrolysate offsite in order to avoid further delay in munitions processing. However, the committee also recognizes the tension that would be created in the decision process if these measures are implemented before they are needed. Making these preparations beforehand should in no way bias the decision in favor of offsite shipment. In the event that offsite shipment needs to be considered, the decision needs to be based on the application of an established decision framework and appropriate consultation with stakeholders.

Finding 7-18. The SCWO system to be used at BGCAPP has been subjected to numerous tests with hydrolysate simulants and appears to be a mature technology. Likewise, the RO system at the heart of the WRS is a proven technology for desalinating water. However, these technologies have not been used with actual hydrolysates in a continuously operating environment for the 3 years during which it is expected to perform at BGCAPP.

Recommendation 7-10. Although the SCWO and WRS appear to be capable of processing hydrolysate at BGCAPP, and a comprehensive preoperational testing program to improve performance will be undertaken, there is still a reasonable possibility that at some point during BGCAPP operations, a decision may need to be made to ship hydrolysate or SCWO effluent offsite. As a precaution, BGCAPP management should prepare for this contingency by taking all necessary actions having long lead times well in advance of such a decision.

REFERENCES

BPBG (Bechtel Parsons Blue Grass). 2013. *Test Report for Supercritical Water Oxidation (SCWO) First-of-a-Kind (FOAK) Test (Preliminary Draft)*. SDN-24915-00-GQY-GGEN-00094. Richmond, Ky.

NOBLIS. 2008. *Offsite Disposal of ACWA Hydrolysates*. NTR 2008-61129. Falls Church, Va.

NRC (National Research Council). 2008. *Review of Secondary Waste Disposal Planning for the Blue Grass and Pueblo Chemical Agent Destruction Pilot Plants*. Washington, D.C.: The National Academies Press.

NRC. 2012. *Letter Report: The Blue Grass Chemical Agent Destruction Pilot Plant's Water Recovery System*. Washington, D.C.: The National Academies Press.

NRC. 2013. *Assessment of Supercritical Water Oxidation System Testing for the Blue Grass Chemical Agent Destruction Pilot Plant*. Washington, D.C.: The National Academies Press.

1

Introduction

In 1993, the United States signed the Chemical Weapons Convention (CWC), an international treaty outlawing the production, stockpiling, and use of chemical weapons.[1] The chemical weapons stockpiles at five of the U.S. chemical weapons storage sites have now been destroyed. At those sites, the munitions were robotically opened and the chemical agent was removed, collected, and incinerated.[2] Chemical agent stored in bulk containers at two other chemical storage sites have been destroyed by hydrolyzing the agent with either hot water or caustic.[3]

The remaining two sites with chemical weapons stockpiles are the Pueblo Chemical Depot (PCD) near Pueblo, Colorado, and the Blue Grass Army Depot (BGAD) near Richmond, Kentucky. The choice of technology to destroy chemical weapons at these sites emerged out of a complex history. As discussed in Chapter 3, in the event that currently planned treatment processes underperform or fail, this history will influence the way that stakeholders approach any consideration of alternatives. The early history of the chemical weapons destruction program (described in more detail in Appendix A) was characterized by significant controversy and social distrust about the Army's preference for incineration to dispose of the stockpile. Much of this controversy had its origins in the questions raised by residents living nearby PCD and BGAD.

A dramatic culmination in the dynamic of growing social distrust and Army insistence on incineration occurred in 1997. A number of factors led to this shift, including (1) mounting pressure on the Army from Congress and (2) in local communities in opposition to incineration to achieve chemical demilitarization (Durant, 2007; Futrell and Futrell, 2012; GAO, 1995a and 1995b). In 1996, Congress enacted laws that froze funds for construction of baseline incineration facilities at PCD and BGAD.[4] These laws further directed the Army to identify and evaluate at least two alternative technologies for the chemical agent destruction. The Assembled Chemical Weapons Assessment (ACWA) program was established to demonstrate other means of destroying the chemical agent.

Congress also required a more robust effort at public involvement by ACWA, which led to the initiation of the 5-year-long dialogue on assembled chemical weapons assessment in 1997 (Goldberg, 2003; Futrell, 2003; and Keystone Policy Center, 2004), referred to from here on as the Dialogue Group. Input was sought from a broad range of interested and affected parties about technical and social criteria for comparing alternative technologies, assessment of the alternative technologies using these criteria, and identification of sites appropriate for the implementation of the alternative technologies. The Dialogue Group included representatives of local citizens; federal, state, and local regulators; the Army; and the National Research Council (NRC). The NRC also conducted reviews of candidate nonincineration technologies for treatment of the chemical stockpiles in Colorado and Kentucky.

Published accounts of the Dialogue Group highlight its importance for rallying public support, enhancing ACWA responsiveness to public concerns and preferences, building the capacity of local stakeholders to participate in highly technical discussions, and rebuilding trust (Goldberg, 2003; Futrell, 2003; and Futrell and Futrell, 2012). After the technologies had been selected, the Dialogue Group was disbanded and Citizens' Advisory Commissions (CACs) were established in Colorado and Kentucky. Both CACs include members of the Dialogue Group. The ACWA program has resulted in the selection of alternative technologies at the two sites and is now the Program Executive Office (PEO)

[1] Convention on the Prohibition of the Development, Production, Stockpiling and Use of Chemical Weapons and on their Destruction. The treaty entered into force in 1997.

[2] These sites were located on Johnston Atoll and in Anniston, Alabama; Pine Bluff, Arkansas; Tooele, Utah, and Umatilla, Oregon.

[3] These sites were located at Aberdeen Proving Ground, Maryland, and Newport, Indiana.

[4] Public Laws 104-201 and 104-208.

for Assembled Chemical Weapons Alternatives, still known by the acronym ACWA.

This report pertains to the destruction of chemical weapons containing nerve agent currently stored at BGAD, and also of energetic materials contained within these munitions. The nerve agent stockpile at BGAD consists of GB (sarin) and VX, in varying munition configurations, including 115-mm M55 rockets, 155-mm projectiles, and 8-in. projectiles. There is also an inventory of chemical munitions containing mustard agent at BGAD. The mustard stockpile will be disposed of using an explosive destruction technology in a separate facility at BGAD. The energetics in the M55 rocket consist of a burster (Comp. B: 60 percent RDX and 40 percent TNT) and M28 propellant (primarily 60 percent nitrocellulose and 23.8 percent nitroglycerine). The nerve-agent-containing projectiles at BGAD do not contain energetics.[5]

The facility being constructed to destroy the nerve agent munitions at BGAD is called the Blue Grass Chemical Agent Destruction Pilot Plant (BGCAPP). At the writing of this report, main plant construction is over 90 percent complete. The process of preparing the plant and its employees for munitions destruction is called systemization. According to ACWA, "Systemization is the testing of all process components, subsystems, and systems, and the demonstration the plant, procedures, and personnel are ready for toxic operations." (BPBGT, 2012, p. 20)

Some parts of the plant have been in various phases of systemization while systemization for other parts of the plant has yet to be started. At the writing of this report, main plant systemization was approximately 30 percent complete. Processing of the nerve agent munitions through the plant is scheduled to begin in the second quarter of FY2018. BGCAPP expects that it will require approximately 3 years to completely destroy the nerve agent in the BGAD stockpile and complete secondary waste processing, with operations scheduled to be completed in the second quarter of FY2021.[6]

The GB and VX will be destroyed by caustic (sodium hydroxide, NaOH) hydrolysis. Energetic materials contained within the chemical munitions will also be treated by caustic hydrolysis. The hydrolysis process results in a waste residual referred to as hydrolysate. BGCAPP intends to blend the agent and energetics hydrolysates and then treat the combined hydrolysate in supercritical water oxidation (SCWO) reactors. There will be a GB phase and a VX phase of operations; the GB and VX hydrolysates will be blended with energetics hydrolysate, but GB and VX hydrolysates will not be blended with each other. The nerve agent hydrolysates, blended with energetics hydrolysate, will go through SCWO separately. The SCWO effluent then will be processed through a water recovery system (WRS), the heart of which is a reverse osmosis unit. Recovered water will be recycled to the SCWO process as quench water in lieu of drawing makeup water. The BGCAPP hydrolysis treatment processes are described in Chapter 2 of this report.

While destruction of the nerve agents is conducted under the auspices of the CWC, because the hydrolysate contains Schedule 1 and 2 compounds under the CWC, the SCWO process is also subject to CWC oversight.[7] The Schedule 1 and 2 compounds in the hydrolysates are identified in Chapter 2.

STATEMENT OF TASK

The project is to be undertaken in two reports: a first report, on the Pueblo Chemical Agent Destruction Pilot Plant (PCAPP), to be delivered 9 months from start of contract, and a second report, on the Blue Grass Chemical Destruction Agent Pilot Plant (BGCAPP), to be delivered at the end of the 15-month contract period.

The National Research Council will establish an ad hoc committee to consider the following study objectives for the BGCAPP second report:

- Develop criteria for successfully treating the hydrolysate in the designed SCWO and Water Recovery Systems;
- Identify systemization data that should factor into the criteria/decision process;
- If the present treatment is not satisfactory, identify potential modifications that would allow continued onsite processing, e.g., additional buffer storage or shipment of excess hydrolysate beyond onsite capacity;
- Identify the downstream impacts to plant operations if offsite shipment is required;
- Determine if aluminum recovery from energetic hydrolysate prior to shipment offsite will be necessary;
- Determine if separate waste disposal options are required for agent and energetics hydrolysates;
- Identify any regulatory requirements for offsite hydrolysate shipment and treatment options, and evaluate transportation risks that could be expected; and
- Consider stakeholder interests and solicit stakeholder input.

---

[5] J. Barton, BGCAPP chief scientist, Battelle, "BGCAPP Agent and Energetics Treatment Processes," presentation to the committee on January 27, 2015.

[6] R. Goetz, assistant project manager, Parsons, BGCAPP, "Project Overview," presentation to the committee on January 27, 2015.

[7] Under the Chemical Weapons Convention, Annex on Chemicals, Schedule 1 chemicals are those that were developed, produced, stockpiled, or used as a chemical weapon, or are chemicals that would pose a high risk to the object and purpose of the Convention by virtue of their high potential for use as a chemical weapon. Schedule 2 are those chemicals that pose a significant risk to the object and purpose of the Convention because they possess such lethal or incapacitating toxicity and other properties that could enable them to be used as a chemical weapon, or may be used as a precursor in one of the chemical reactions at the final stage of formation of or is used in the production of a Schedule 1 or a Schedule 2 chemical, https://www.opcw.org/chemical-weapons-convention/annexes/annex-on-chemicals/.

OVERVIEW OF THE COMMITTEE'S WORK AND THE FACTORS IMPACTING ON IT

This committee has done its work in two distinct parts as broken down by the statement of task. The first part addressed the Pueblo Chemical Agent Destruction Pilot Plant (PCAPP), while this report, the second part, addresses the hydrolysate treatment processes at BGCAPP. The composition of the committee changed slightly between the two parts. The committee that completed the PCAPP report is referred to as the PCAPP hydrolysate committee. The committee that completed the work in this report is referred to as the BGCAPP hydrolysate committee (or, simply, the committee).

The complete statement of task, above, required assessment of hydrolysate treatment processes at both PCAPP and BGCAPP. Construction of the PCAPP plant was anticipated to have been completed by the time the BGCAPP hydrolysate committee was to begin its deliberations and was scheduled to begin agent operations in September 2015.[8] It was therefore determined that it would be most beneficial for the NRC to first address PCAPP. Hence, the report prepared by the PCAPP hydrolysate committee was nearing completion and publication as the effort for BGCAPP was beginning.

The first BGCAPP hydrolysate committee meeting took place in January 2015 and the PCAPP report was released in March 2015. While there are a number of important differences between PCAPP and BGCAPP, including the agents being destroyed and the treatment methods, the main thrusts of both reports were the same:

- Establish criteria for successfully treating the hydrolysate,
- Identify systemization data that may factor into the criteria and the decision process,
- Suggest potential modifications that would allow continued onsite processing of the hydrolysate should the hydrolysate treatment systems develop issues, and
- Evaluate the potential for offsite transport of hydrolysate should the hydrolysate treatment processes underperform or fail.

This report follows the underlying structure and concepts used in the PCAPP report. The committee determined that having the opportunity to consider important concerns expressed by PCAPP stakeholders regarding the PCAPP report would in fact be beneficial in writing the BGCAPP report. Therefore, the committee determined, with input from the NRC, that feedback from the PCAPP report would be considered in writing the BGCAPP report.

[8] The NRC committee learned at a meeting of the NRC's Committee on Chemical Demilitarization held on February 18, 2015, in Washington D.C., that initiation of agent destruction at PCAPP has been delayed to the beginning of January 2016.

INTRODUCTION TO BGCAPP HYDROLYSATE TREATMENT TECHNOLOGIES

As indicated in the NRC's 2013 report on SCWO, SCWO is considered first-of-a-kind (FOAK) technology (NRC, 2013). SCWO reactors at BGCAPP will subject blended agent and energetics hydrolysate to very high temperatures and pressures, resulting in the destruction of organic materials within the hydrolysate to carbon dioxide, water, and salts. The total organic carbon content of the water that will be released from the SCWO reactors is anticipated to be less than 10 parts per million.[9]

The chemical agent will be treated to 99.9999 percent destruction in agent concentration before entering the SCWO process.[10] The hydrolysis process breaks up large complex molecules into smaller ones, destroying the nerve agents and eliminating their acute toxicity. Thus, the hydrolysate should no longer contain GB or VX. The level of destruction efficiency was first established by the Kentucky legislature and is a regulatory requirement. Details of how the regulatory process impacts decisions at BGCAPP are provided in Chapter 4.

Very little, if any, agent will remain in the hydrolysate. However, because of the high salt content of the hydrolysate, including a number of different salt species, the blended agent and energetics hydrolysate feed presents a challenge to SCWO. Considering the nature of the hydrolysate and issues with corrosion and salt buildup, extensive FOAK testing was conducted prior to the units being shipped to BGCAPP. This FOAK testing was reviewed by an NRC SCWO committee in 2013, and while that committee had many findings and recommendations for BGCAPP, it nevertheless concluded that "SCWO is a mature technology with a strong scientific and engineering base underlying it"[11] (NRC, 2013). For example, SCWO solutions have been designed for the treatment of diverse waste materials, water purification, the recovery of precious metals from catalytic materials, and the production heat and power. The NRC 2013 concluded that while extensive testing of these systems has been conducted in a laboratory setting, the process has never been operated at full scale on the chemical feeds it will process at BGCAPP. This committee agrees with that conclusion.

In addition, ACWA had commissioned an earlier NRC report, in 2012, to evaluate the BGCAPP WRS (NRC, 2012). That committee identified a number of issues in its report, but, overall, it had no overarching concerns that the WRS process would not work on BGCAPP SCWO effluent if its recommendations are followed.[12] WRS issues are discussed in Chapters 6 and 7.

[9] G. Lucier, deputy chief scientist, BGCAPP, D. Linkenheld, SCWO start-up supervisor, and L. Austin, waste manager, BGCAPP, "SCWO Process: Cradle to Grave," presentation to committee on January 28, 2015.

[10] 401 KAR 34:350.

[11] Two of the 2013 SCWO committee members are also members of this BGCAPP hydrolysate committee.

[12] Two of the 2012 WRS committee members are also members of this BGCAPP hydrolysate committee.

The committee received 2 days of briefings from ACWA and BGCAPP staff, from public stakeholders, and from Kentucky regulators during the January 2015 meeting at BGCAPP. The committee has since reviewed extensive documentation on the BGCAPP process, including the earlier 2013 and 2012 NRC reports, and additional documentation provided by ACWA and BGCAPP. As discussed in Chapters 6 and 7 of this report, however, there are a number of technical concerns with operation of the SCWO and the WRS. BGCAPP plans to conduct extensive preoperational testing concurrent with facility systemization that will evaluate these concerns. This testing is expected to result in facility changes (e.g., equipment and operating procedures) that will increase the likelihood of success. Technological underperformance or failure of the SCWO and/or WRS could result in BGCAPP not meeting its overall performance criteria, as set forth in Chapter 6, including not only efficient and effective operations but also regulatory and treaty compliance, public expectations, and an established destruction schedule.

Finding 1-1. It is expected that extensive preoperational testing of the SCWO and the WRS will be performed concurrently with systemization to help reduce the uncertainty in expected technological performance. However, considering the first-of-a-kind nature of these technologies and the need for them to properly function in tandem, there is a possibility that technological issues with these systems may prevent BGCAPP from meeting all of its overall performance criteria.

As mentioned above, there are a variety of reasons that the SCWO or WRS systems might not perform satisfactorily. Chapter 7 provides a detailed discussion of the technical factors that could lead to underperformance of these treatment systems. For example, corrosion and salt clogging could lead to frequent repairs that could shut down hydrolysate treatment in the SCWO units. Occurrences like these could also lead to other issues, such as storage of the hydrolysate for unanticipated times, and idle periods for BGCAPP workers. More importantly, however, if these or similar problems occur, the actual munitions disassembly and hydrolysate production may have to be interrupted until the downstream processes are able to catch up. In the event that one or more of the hydrolysate treatment systems are shut down, and as hydrolysate storage nears capacity, destruction of the primary stockpile at BGAD may need to be halted unless there is an alternative means for treating the hydrolysate.

A delay in the disposal of munitions and agent at BGCAPP is extremely undesirable. The nerve agent munitions to be treated at the BGCAPP have been stored at the BGAD for over 50 years, representing a steady-state or even increasing risk profile for the community (due, for example, to the potential for the munitions to spring new leaks); final destruction of these munitions at BGAD will eliminate this risk. Delays in the destruction process will halt this risk reduction and protract the risk that the community faces. If the SCWO or WRS processes do not perform as expected, it would be necessary for any decisions that might delay BGCAPP destruction operations to consider potential impacts beyond the plant, such as the risk associated with storing aging chemical weapons for longer periods of time, public expectations, regulatory and treaty issues, and employee reassignment, furloughs, or layoffs. The committee firmly believes that destruction of the stockpile at BGAD must continue, because it is destruction of the munitions and the agents that will eliminate the primary risk to the community. Hence, the committee believes that it is necessary to establish a backup plan—an alternative to the onsite hydrolysate treatment processes.

Recommendation 1-1. Considering that the SCWO and the WRS may not perform satisfactorily and that this underperformance could result in delays in the destruction of chemical agent at BGCAPP, increasing the primary risk to the community associated with continued munitions storage, BGCAPP should establish a backup plan as an alternative to the onsite hydrolysate treatment processes.

One alternative for treatment of the hydrolysate that might be more quickly implementable than other alternatives would be to ship the hydrolysate and/or SCWO effluent offsite to an existing, prequalified, and fully permitted treatment, storage, and disposal facility. The decision to implement offsite shipment would impact the downstream plant units that were constructed to treat the hydrolysate generated from the agent and energetics hydrolysis processes. Workers associated with the discontinued SCWO or WRS processes might need to be reassigned and retrained, placed on furlough, or laid off, resulting in the loss of specially trained personnel. This is not the desire of ACWA program staff, the BGCAPP contractors, or the local stakeholders.

The choices faced with possible underperformance of the SCWO or WRS are highly complex. All alternatives will entail interconnected considerations about technological capabilities and limitations, in conjunction with ACWA commitments, contractual requirements, public stakeholder opinion, regulatory permitting requirements, schedule delays, and cost.

To study this situation at BGCAPP, ACWA requested that the NRC form an ad hoc committee, the Committee on Review Criteria for Successful Treatment of Hydrolysate at the Pueblo and Blue Grass Chemical Agent Destruction Pilot Plants, to assess the SCWO and WRS processes, develop criteria by which to judge success, consider onsite mitigation strategies should the SCWO or WRS systems underperform, and explore the potential for offsite transport of the hydrolysate.

CRITERIA FOR SUCCESSFULLY TREATING HYDROLYSATE

Note that the statement of task, above, calls on the committee to, among other things, "develop criteria for successfully treating the hydrolysate in the designed SCWO and Water Recovery Systems." Chapter 6 lays out two categories of criteria:

- *Performance requirements.* Conditions that must be met under regulatory permits and under CWC treaty obligations, and
- *Performance goals.* Primarily oriented toward process performance and schedule and should be met to enable satisfactory system performance.

Performance requirements are crucial in achieving successful operation of the BGCAPP. If these requirements cannot be met, and if the time it takes to destroy the actual munitions increases as a result of the degraded performance, risk reduction goals associated with destruction of the stockpile will not be achieved in a timely manner, and consideration of the offsite hydrolysate and/or SCWO effluent option becomes more likely. Performance goals, in comparison, represent goals for the treatment process that should be met to achieve satisfactory performance, but nonattainment of one or more of these goals will typically not result, at least immediately, in consideration of the offsite option.

The committee did not develop new success criteria for the SCWO and WRS, as called for in the statement of task. While criteria are presented in Chapter 6, they were not "developed" by the committee. Rather, these criteria and goals were garnered from a review of regulatory requirements and BGCAPP documentation. The committee believes the criteria presented in Chapter 6 are reasonable for this stage of the project because they are based on prior SCWO testing with both simulants and actual hydrolysates and on SCWO performance modelling. The committee feels strongly that any new criteria should be developed based on preoperational testing, which has yet to be initiated. The committee did not want to lock BGCAPP into new criteria developed without the benefit of preoperational testing and systemization data, and was especially concerned that new criteria it developed could be incorporated into permit documentation. Hence, while Chapter 6 establishes performance requirements and goals, no new criteria were developed by the committee.

CONSIDERATION OF ALTERNATIVE TECHNOLOGIES

In the course of its interactions with public stakeholders, the committee received an inquiry from the Kentucky CAC about alternative technologies and approaches that could be employed onsite should the SCWO or WRS systems underperform or fail altogether. Specifically, as indicated earlier, BGAD intends to use an explosive destruction technology—that is, the Static Detonation Chamber (SDC)—to destroy its entire mustard munition stockpile. The committee was asked following its January 2015 meeting if this technology could be used to dispose of the nerve agent and energetics hydrolysates.[13]

The statement of task does not call for the committee to evaluate technologies other than SCWO and the WRS. Therefore, the committee did not conduct an evaluation of other possible onsite technologies for the treatment or disposal of hydrolysate.

It is, nevertheless, possible to make some general predictions in connection with adopting the SDC technology to replace SCWO for the onsite treatment of hydrolysate. The SDC is a double-walled detonation chamber designed to destroy munitions containing chemical agent. The heating of the explosives in the munitions and/or the pressure generated from the heated liquid agent contents eventually cause the munitions to rupture and the agent to be destroyed in the >1,000°F operating temperature in the chamber.

Thus, the SDC is designed to treat an entire munition that contains agent rather than a bulk liquid material such as agent-derived hydrolysate. There are a number of other technologies that were considered for treatment of the hydrolysate in the course of selecting SCWO for BGCAPP. However, as indicated previously, the technologies selected for BGCAPP, specifically hydrolysis followed by SCWO, arose out of a complex history involving many stakeholders; many different technologies were evaluated during this process, culminating in the selection of hydrolysis and SCWO. Further, any effort to adopt another technology would necessitate a formal evaluation effort to select the most appropriate technology. Considering, as stated previously, that continued storage of the nerve agent munitions represents an ongoing risk to the community, processes to select an alternative technology, develop designs and equipment, construct and test new facilities, and obtain permits and other regulatory approvals, would significantly prolong this risk. In addition, further delays would contravene CWC treaty obligations.

In the event that the SCWO and WRS underperform or fail altogether, the most likely option that would help reduce the risk to the community in a timely manner and in compliance with regulatory requirements is offsite transport of the hydrolysate. Future consideration of the full range of options, if contemplated by ACWA, CAC, and the Kentucky Chemical Destruction Community Advisory Board, would weigh these issues very carefully. While a comprehensive and detailed assessment of adopting another technology to replace SCWO for the onsite treatment of hydrolysate is beyond the scope of charge to the committee, the discussions of the decision framework (Chapter 6), public involvement (Chapter 3), and

[13] E-mail from Douglas Hindman, chair, Kentucky Citizens' Advisory Commission, to committee on February 3, 2015.

permitting (Chapter 4) would apply to any alternatives that might be considered.

The committee notes also that the safety of hydrolysate treatment operations is a concern. The committee, however, was not tasked to look into safety. It was apparent, from ACWA and BGCAPP presentations during the January 2015 briefings in Kentucky, as well as the committee tour of the BGCAPP facility, that safety was considered to be of very high importance. The committee believes that safety must remain the highest priority through all operations, including, if necessary, offsite transport of hydrolysate.

Finally, the costs of modifying the SCWO and the WRS at BGCAPP to improve their performance, should this be needed, must be taken into consideration by BGCAPP. If additional funding is needed to modify onsite operations or to allow offsite transportation of hydrolysate, significant delays to the BGCAPP operating schedule and increases in its life-cycle costs may result. Hence, these costs must be carefully considered. While the committee recognizes this situation, it was not tasked to conduct cost analyses of various mitigation activities. Consequently, although cost goals are recognized in this report, they are not explicitly evaluated.

ORGANIZATION OF THIS REPORT

Chapter 2 of this report provides information on the composition of the hydrolysates that will be produced by BGCAPP and describes in more detail the processes planned for use in treating them. Chapter 3 discusses the history of public stakeholder involvement regarding the potential for offsite shipment and treatment of hydrolysate and current concerns related to this. Chapter 4 reviews regulatory considerations at the federal, state, and local levels for onsite process modifications and possible offsite shipment and treatment of hydrolysate, and addresses the requirements of the CWC in this regard. Chapter 5 discusses Department of Transportation regulations for the shipment of hazardous material, presents historical data that quantify the risk of hazardous material shipment, and identifies risks associated with the offsite shipment of hydrolysate, including historical experience with the shipment of agent hydrolysates. Chapter 6 establishes criteria for successfully treating the BGCAPP hydrolysates and identifies systemization data that should factor into the criteria and decisions regarding onsite mitigation approaches and possible offsite transport and disposal of the hydrolysate. Chapter 7 discusses underperformance and failure risks and contingency options as well as the downstream impacts of a decision to ship hydrolysate and/or SCWO effluent offsite. Chapter 7 also addresses issues with the potential for aluminum to be present in the hydrolysate due to dissolution of aluminum M55 rocket parts and also whether the combination of energetics and agent hydrolysate might impact a decision to send combined hydrolysate offsite.

There are also four appendixes. Appendix A provides a chronology of events at BGAD to the present, with focus on public involvement. Appendix B contains material documenting committee efforts to encourage public involvement in this study and material received from the CAC/CDCAB. Appendix C provides biographical sketches of committee members. Appendix D identifies the committee meetings and locations.

REFERENCES

BPBGT (Bechtel Parsons Blue Grass Team). 2012. *Systemization Implementation Plan (SIP)*, Rev. 0, August 30. Richmond, Ky.: Blue Grass Chemical Agent Destruction Pilot Plant Program Office.

Durant, R.F. 2007. *The Greening of the U.S. Military*. Washington, D.C.: Georgetown University Press.

Futrell, R., and D. Futrell. 2012. Expertise and alliances: How Kentuckians transformed the US Chemical Weapons Disposal Program. S. McSpirit, L. Faltraco, and C. Bailey, eds., Chapter 9 in *Confronting Ecological Crisis in Appalachia and the South,* pp. 171-193. Lexington, Ky.: University Press of Kentucky.

Futrell, R. 2003. Technical adversarialism and participatory collaboration in the U.S. chemical weapons disposal program. *Science, Technology, & Human Values* 28(4): 451-482.

GAO (General Accounting Office). 1995a. *Chemical Stockpile Disposal Program Review*. NSIAD-95-66R. Washington, D.C.: General Accounting Office.

GAO. 1995b. *Chemical Weapons Disposal: Issues Related to DOD's Management*. T/NSIAD-95-18. Washington, D.C.: General Accounting Office.

Goldberg, M. 2003. Strengthening the link between project planning and environmental impact assessment: The assembled chemical weapons assessment dialogue process. *Environmental Practice* 5(4): 313-320.

Keystone Policy Center. 2004. ACWA Dialogue Close-out Report (draft), https://www.keystone.org/.

NRC (National Research Council). 2012. *Letter Report: The Blue Grass Chemical Agent Destruction Pilot Plant's Water Recovery System.* Washington, D.C.: The National Academies Press.

NRC. 2013. *Assessment of Supercritical Water Oxidation System Testing for the Blue Grass Chemical Agent Destruction Pilot Plant.* Washington, D.C.: The National Academies Press.

2

The Supercritical Water Oxidation and Water Recovery System Processes Planned for Use at the Blue Grass Chemical Agent Destruction Pilot Plant

This chapter provides background information about the process planned for use at the Blue Grass Chemical Agent Destruction Pilot Plant (BGCAPP) to dispose of the nerve agents and energetic materials contained in the nerve agent munitions stored at the Blue Grass Army Depot (BGAD). Subsections provide an overview of the entire process, including a discussion of the production and characterization of the hydrolysates, a description of the supercritical water oxidation (SCWO) system, and a description of the water recovery system (WRS). Two earlier National Research Council (NRC) reports also provide background information about the SCWO system (NRC, 2013) and the WRS (NRC, 2012) and may be of interest to readers of this report.

BRIEF DESCRIPTION OF THE BGCAPP PROCESS

The energetic materials and the nerve agents stored in munitions at BGAD will be removed from the projectiles and rockets that contain them in a munitions pretreatment area. The energetic materials in the munitions include RDX ($C_4H_8N_8O_8$), HMX ($C_3H_6N_6O_6$), TNT ($C_7H_5N_3O_6$), nitroglycerin ($C_3H_5N_3O_9$), nitrocellulose, and tetryl ($C_7H_5N_5O_8$). The chemical agents include GB ($C_4H_{10}FO_2P$) and VX ($C_{11}H_{26}NO_2PS$). Munitions containing H ($C_4H_8Cl_2S$) are also stored at BGAD. These will be destroyed by a separate explosive destruction technology.

These materials will then be neutralized separately by high-temperature caustic hydrolysis. This process breaks up large complex molecules into smaller ones, destroying the nerve agents and eliminating their acute toxicity. Thus, the hydrolysate should no longer contain GB or VX. The drained projectile bodies, along with any material adhering to them, will be thermally treated in the metal parts treater at 1,000°F for 15 minutes. The offgas from the metal parts treater (MPT) goes to a thermal oxidizer unit, which operates at approximately 2,200°F and has a 3-sec residence time, to destroy any trace organic materials. The rocket warheads and bursters will be hydrolyzed in the energetics batch hydrolyzers (EBHs). Residual metal parts will also be sent to the MPT. The metal from the MPT will go offsite for disposal or recycling.

After laboratory analysis to ensure the hydrolysates meet release requirements, including that agent has been destroyed to regulatory requirements, the VX, GB, and energetics hydrolysates will be stored separately outdoors in closed tanks. The energetics hydrolysate will be diluted and acidified to facilitate the subsequent removal of aluminum in the aluminum filtration system, as discussed later in this chapter. The GB and VX hydrolysates will then be individually blended with energetics hydrolysate, combined with additives (e.g., HCl, H_2SO_4, NaCl, S) to improve SCWO processability, and then, along with fuel and compressed air, be fed to a SCWO reactor for the purpose of converting the organic carbon to CO_2. The air feed provides the oxygen for oxidation, and the fuel (isopropyl alcohol) is used to maintain the reactor temperature. Additional SCWO feed streams include spent decontamination solution, condensate from the metal parts treater offgas treatment system, and other condensates. Sufficient mixing of the various components in the SCWO feed stream should help to ensure that a homogeneous solution enters the SCWO reactor. Quench water is added at the bottom of the SCWO reactors to cool the SCWO effluent and to keep salts in solution downstream of the SCWO reactors. The SCWO facility consists of three identical SCWO trains. Each SCWO train consists of a feed module and a reactor module. The processing of the hydrolysates with SCWO is required to comply with regulatory permits as well as to meet requirements of the Chemical Weapons Convention (CWC). The treatment process is shown in the block flow diagram of Figure 2-1.

After being cooled, the SCWO reactor effluent will be separated into gas and liquid phases. The gas phase passes through carbon filters before being released to the atmosphere. If the composition of the liquid phase meets treatment specifications, it proceeds to the SCWO effluent tank and is then treated in the WRS to produce a brine stream for

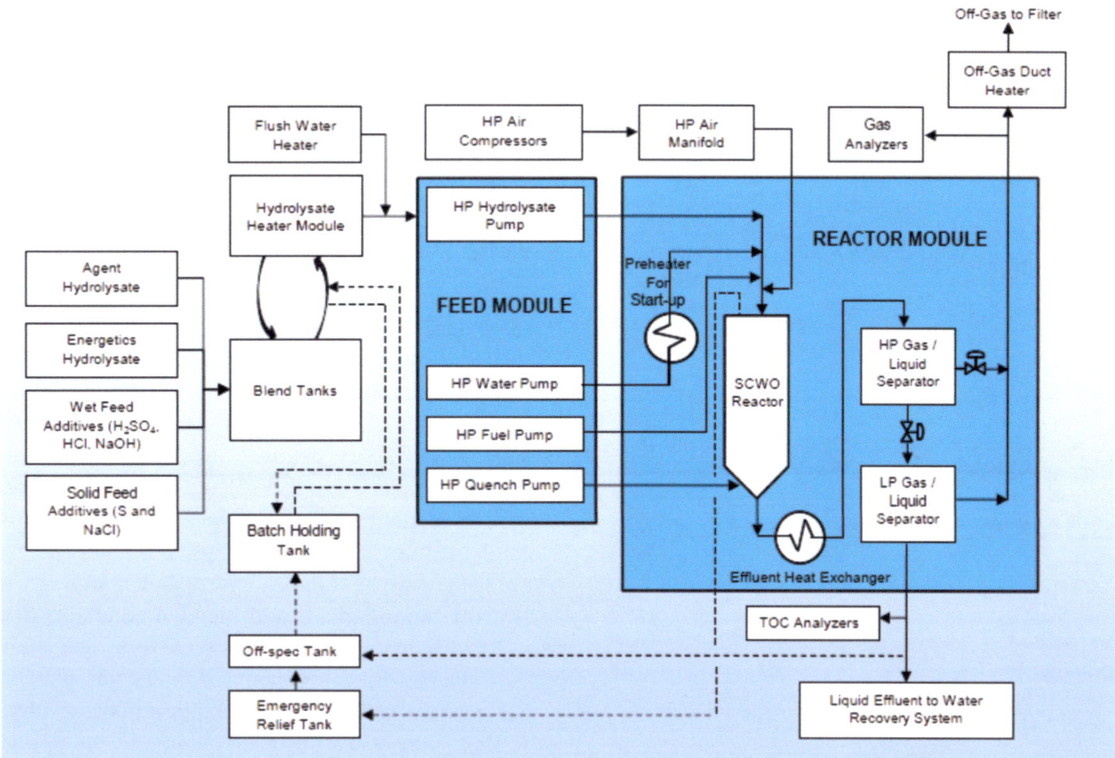

FIGURE 2-1 Schematic diagram of SCWO reactors. SOURCE: George Lucier, deputy chief scientist, BGCAPP; Dave Linkenheld, SCWO start-up supervisor; and Larry Austin, waste manager, BGCAPP, "SCWO Process: Cradle to Grave," presentation to the committee on January 28, 2015.

ultimate offsite disposal and a product water stream that will be recycled as quench water for the SCWO reactors. Off-spec SCWO effluent will go to the off-spec tank, be blended with hydrolysate, and go through the reactor again until the treatment specifications are met.

CHARACTERIZATION OF AGENT AND ENERGETICS HYDROLYSATES

This section provides information about the likely compositions of the three different hydrolysates. The information comes from an understanding of the chemistry that occurs during hydrolysis and from process simulations performed by BGCAPP. The total volume of hydrolysate anticipated to be generated at BGCAPP is approximately 2.5 million gallons, with about 7 percent of this volume coming from VX neutralization, 37 percent from GB, and the remaining 56 percent from energetics.

The actual compositions of the hydrolysates that will be produced at BGCAPP are not yet known with certainty, because no detailed, quantitative, analytical characterization of actual hydrolysate will be available until the plant begins to generate hydrolysate. Moreover, the variability in the hydrolysates' compositions due to variability in the compositions of the agent and energetics themselves is not known. Nevertheless, enough was known about the anticipated compositions of these streams to permit formulation of the simulated hydrolysates used in the SCWO first-of-a-kind (FOAK) testing.

Finding 2-1. The compositions of the hydrolysates and their batch-to-batch variability have only been estimated at this point. No detailed, quantitative analysis is available.

GB Hydrolysate Composition

Nerve agent GB will be neutralized in caustic at 160°F for 4 hr. The chemistry of GB neutralization has been well documented and is well known. The main products are sodium isopropyl methylphosphonate (IMP) and sodium fluoride (NaF). Tables 2-1, 2-2, and 2-3 provide the compositions of the GB hydrolysate simulant, the energetics hydrolysate simulant, and the blended simulant used in FOAK testing of SCWO. Note that the simulated hydrolysates were not intended to contain all of the chemical species that could be present in actual agent hydrolysate. Rather, they were designed to mimic the solids-forming elements, heating value, and content of key species—tributylamine for GB and organic nitrogen for VX (see next subsection)—expected in the actual agent hydrolysate (BPBG, 2013).

TABLE 2-1 Simulated GB Agent Hydrolysate Used in FOAK Testing

Constituent (7.5 wt% reacted GB) (without additives)	Wt%
Deionized water	87.03
Sodium fluoride (NaF)	1.84
Sodium hydroxide (NaOH)	3.84
Dimethylmethyl phosphonate ($C_3H_9O_3P$) (DMMP)	5.91
100% Isopropanol (C_3H_8O) (IPA)	1.18
Tri-*n*-butylamine ($C_{12}H_{27}N$)	0.20
Total	100.00

SOURCE: BPBG (2013).

TABLE 2-2 Simulated Energetics Hydrolysate Used in GB FOAK Testing

Constituent (with additives)	Wt%
Deionized water	78.14
Sodium chloride (NaCl)	5.30
Sodium sulfate (Na_2SO_4)	12.71
Sodium nitrite ($NaNO_2$)	0.39
Sodium formate ($NaCHO_2$)	1.18
35% Hydrochloric acid (HCl)	1.92
Sulfur (S)	0.36
Total	100.00

SOURCE: BPBG (2013).

TABLE 2-3 Simulated Blended Hydrolysate Used in FOAK Testing for GB[a]

Constituent (with additives)	Wt%[b]
Deionized water	80.64
Sodium sulfate (Na_2SO_4)	9.14
Sodium chloride (NaCl)	3.81
Sodium hydroxide (NaOH)	1.08
Dimethyl methylphosphonate $C_3H_9O_3P$ (DMMP)	1.66
35% Hydrochloric acid (HCl)	1.38
Sodium formate ($NaCHO_2$)	0.85
Sodium fluoride (NaF)	0.52
100% Isopropanol (C_3H_8O) (IPA)	0.33
Sodium nitrite ($NaNO_2$)	0.28
Sulfur (S)	0.26
Tri-*n*-butylamine ($C_{12}H_{27}N$) (TBA)	0.06
Total	100.00

[a] 2.5 parts simulated energetics hydrolysate to 1 part simulated agent hydrolysate.
[b] Agent hydrolysate simulant constituents are diluted further by the additives used in the energetics hydrolysate recipe; hence the dilution factor is 3.558, not 3.5. For DMMP, for example, we have $[(102.33 \times 2.5) + 100]/100 = 3.558$; $5.91/3.558 = 1.66$ wt%.
SOURCE: BPBG (2013).

In addition to the components listed in the tables, actual GB hydrolysate will also contain trace components such as dithiane, diisopropyl methylphosphonate (DIMP), triisopropyl phosphate, N,N′-diisopropyl urea (DIPU), N-hexyl butanamide, dibutylacetamide, and dibutylbutanamide (Malloy et al., 2007).

VX Hydrolysate Composition

Nerve agent VX will be neutralized in caustic at 194°F for 9 hr. The chemistry of VX neutralization has also been well documented and is well known. The main products from caustic hydrolysis of VX are sodium ethyl methylphosphonate and 2-(diisopropylamino)ethane thiol (sodium salt). A secondary pathway (about 10 percent of the VX) leads to ethanol and EA 2192,[1] which then hydrolyze to sodium methylphosphonate and 2-(diisopropylamino)ethane thiol (sodium salt). Tables 2-4, 2-5, and 2-6 provide the compositions of the VX hydrolysate simulant, energetics hydrolysate simulant, and blended simulant, respectively, used in FOAK testing of SCWO. Note that the simulated hydrolysates were not intended to contain all of the chemical species that could be present in actual agent hydrolysate. Rather, they were designed to mimic the solids-forming elements, heating value, and content of key species (i.e., tributylamine for GB and organic nitrogen for VX) expected in the actual agent hydrolysate (BPBG, 2013).

In addition to the components in the simulants listed in the tables above, actual VX hydrolysate also contains traces of many other compounds, the presence of which has been detected but the precise amounts of which are not known. These trace components include the following (Dejarme and Lecakes, 2008):

- [2-(Diisopropylamino)ethyl]-(2-mercaptoethyl) sulfide
- 1,2-Bis-(2-diisopropylaminoethyl) ethane
- Cyclohexanamine, N-cyclohexyl-
- Morpholine, 4-phenyl-
- N-Isopropylethylenediamine
- (Diisopropylamino)ethanol
- 1,2-Bis-(2-diisopropylaminoethyl) ethane
- 1,3-Dicyclohexylcarbodiimide
- 2-(Diisopropylamino)ethanethiol
- 2-Diisopropylaminoethyl ethyl ether
- 2-Diisopropylaminoethyl ethyl sulfide
- Acetamide, N-(3-methyl-2-buten-1-yl)-
- Acetamide, N-cyclohexyl-
- Arsine, triphenyl-
- Bis(diisopropylaminoethyl) disulfide
- Cyclohexanamine
- Ethane, 1,2-bis(methylthio)-

[1] EA-2192 is a decomposition product of VX that is nearly as toxic as VX. It is a Schedule 1A chemical under the CWC.

TABLE 2-4 Simulated VX Agent Hydrolysate Used in FOAK Testing

Constituent (16.6 wt% reacted VX) (without additives)	Wt%
Deionized water	58.18
Sodium hydroxide (NaOH)	6.37
Dimethyl methylphosphonate ($C_3H_9O_3P$) (DMMP)	7.41
85% Diethanolamine ($C_4H_{11}NO_2$) (DEA)	7.76
56% Sodium isethionate ($C_2H_5NaO_4S$) (SI)	15.63
Ethanol (C_2H_6O) (denatured)	4.64
Total	100.00

SOURCE: BPBG (2013).

TABLE 2-5 Simulated Energetics Hydrolysate Used in VX FOAK Testing

Constituent (with additives)	Wt%
Deionized water	80.32
Sodium chloride (NaCl)	4.26
Sodium sulfate (Na_2SO_4)	10.24
Sodium nitrite ($NaNO_2$)	0.32
Sodium formate ($NaCHO_2$)	0.94
35% Hydrochloric acid (HCl)	1.74
93% Sulfuric acid (H_2SO_4)	0.91
Sodium chloride (NaCl)	1.28
Total	100.00

SOURCE: BPBG (2013).

TABLE 2-6 Simulated Blended Hydrolysate Used in VX FOAK Testing[a]

Constituent (with additives)	Wt%[b]
Deionized water	74.18
Sodium sulfate (Na_2SO_4)	7.40
56% Sodium isethionate ($C_2H_5NaO_4S$) (SI)	4.34
Sodium hydroxide (NaOH)	1.77
Sodium chloride (NaCl)	4.00
85% Diethanolamine ($C_4H_{11}NO_2$) (DEA)	2.15
Dimethyl methylphosphonate $C_3H_9O_3P$ (DMMP)	2.06
Ethanol (C_2H_6O) (denatured)	1.29
35% Hydrochloric acid (HCl)	1.25
Sodium formate $NaCHO_2$	0.68
93% Sulfuric acid (H_2SO_4)	0.65
Sodium nitrite ($NaNO_2$)	0.23
Total	100.00

[a]2.5 parts simulated energetics hydrolysate to 1 part simulated agent hydrolysate.
[b]Agent hydrolysate simulant constituents are diluted further by the additives used in the energetics hydrolysate recipe; hence the dilution factor is 3.558, not 3.5. For DMMP, for example, we have $[(102.33 \times 2.5) + 100]/100 = 3.558$; $5.91/3.558 = 1.66$ wt%.
SOURCE: BPBG (2013).

- Ethane, 1-[(2-diisopropylamino)ethylthio]-2-[(2-diisopropylamino)ethyldithio]-
- N,N-Diisopropylformamide
- Urea, 1-(3-chloropropyl)-3-cyclohexyl-
- Urea, N,N′-bis(1-methylethyl)-

Chemical Weapons Convention Treaty Chemicals in Agent Hydrolysate

In addition to the chemicals in the agent hydrolysate simulant recipes and the trace components listed above for VX hydrolysate, the actual GB and VX hydrolysates may contain chemicals regulated by the Chemical Weapons Convention (CWC). This depends on the chemical reaction pathway that occurs during hydrolysis. The CWC identifies three classes of chemicals on the basis of the scale of their use in legitimate, nonweapons applications. Schedule 1 chemicals have no (or few) uses aside from chemical weapons. Schedule 2 chemicals have small-scale applications, and Schedule 3 chemicals have large-scale applications apart from chemical weapons. Chemicals in each schedule are classified as Part A (can be used directly as weapons) or Part B (can be used to manufacture chemical weapons). The agent hydrolysates may contain chemicals that are listed as Schedule 1A (i.e., EA2192). Schedule 2 compounds that might be present include ethyl methyl phosphonic acid (EMPA), methyl phosphonic acid (MPA), O-ethyl methyl phosphonothioic acid (EMPSH), and isopropyl methyl phosphonic acid (IMPA).

Energetics Hydrolysate Composition

In addition to the nerve agents GB and VX, the M55 rockets in the BGAD stockpile also contain energetic materials, i.e., bursters and propellant from rocket motors. This material will be physically separated during the dismantling of the rockets and will be neutralized at 240-300°F in the EBHs.[2] The chemistry of the hydrolysis step and the resultant composition of the hydrolysate have been well documented. Representative compositions of energetics hydrolysate were given above in Tables 2-2 and 2-5. The energetic materials are expected to form C_1 compounds such as formate and formaldehyde, along with N_2O, NH_3, and N_2. One of the other by-products of energetics hydrolysis is cyanide. BGCAPP has established specific hydrolysis reactor conditions that are designed to significantly reduce the presence of cyanide in the hydrolysate before it is transferred to the SCWO facility.[3] Other components likely to be present in the energetics hydrolysate include acetate, aluminum, ammonia,

[2] Only propellant from rocket motors contaminated with agent will be processed through the EBHs.

[3] J. Barton, chief scientist, BPBG, "Cyanide Mitigation and Worker Protection," presentation to the Kentucky Chemical Demilitarization Citizens' Advisory Commission and Chemical Destruction Community Advisory Board on May 7, 2014, http://www.slideshare.net/acwanews/cyanide-mitigation-and-worker-protection-may-7-2014.

beryllium, calcium, chloride, chromium, cobalt, copper, fluoride, formate, HMX, iron, lead, magnesium, manganese, mercury, molybdenum, nickel, nitrate-N, nitrite-N, phosphorus, potassium, silver, sodium, sulfate, TNT, and zinc (Bonnett and Elmasri, 2002).

The rocket shipping and firing tubes (SFTs) contain polychlorinated biphenyls (PCBs). The PCBs were used as a lubricant to make it easier to slide the rockets into the SFTs.[4] PCB concentrations in the SFTs range from under 50 parts per million (ppm) to more than 2,000 ppm. The SFTs may exit the BGCAPP processes with the rocket motors when separated from the nonleaking agent-filled warheads to be treated or disposed of at a permitted, offsite Toxic Substances Control Act facility. For approximately 200 "leaker" rockets, the rocket motors will be processed through the EBHs, and the hydrolysates from these EBH batches can be expected to contain PCBs leached from the SFTs. BGCAPP calculated that a conservative estimate of the PCB concentrations in each batch of energetics hydrolysate would be 44 ppm.[5]

Dissolved aluminum is present in the energetics hydrolysate because both the M56 warhead and the fuze in an M55 rocket are made of mostly aluminum. During energetics hydrolysis some of this aluminum will react with water to form soluble aluminum species. This aluminum is particularly problematic for the SCWO process, as past studies have shown that aluminum forms solid precipitates in the SCWO reactor that interfere with the flow of materials through the reactor. As a result, most of the aluminum must be removed from the alkaline energetics hydrolysate before it can be blended with agent hydrolysate and treated by SCWO. The energetics hydrolysate will therefore be treated by adding acid to precipitate the aluminum as aluminum hydroxide, after which the slurry will be dewatered using a filter press. The filter cake will then be collected and sent for offsite disposal. The liquid filtrate from the belt press will be collected in storage tanks where it will be kept for blending with agent hydrolysate prior to being fed to the SCWO.

GENERAL DESCRIPTION OF SUPERCRITICAL WATER OXIDATION

SCWO has been studied for nearly 40 years. Several companies have developed technologies to use SCWO for different applications, including the treatment of diverse waste materials, water purification on long-term manned space missions, recovering precious metals from catalytic materials, and producing heat and power (e.g., SCWO of coal–water slurries). This extensive experience with SCWO has revealed that it can be very effective for a wide range of feed materials and that the main long-term operating challenges relate to corrosion and the management of salts in the reactor.

The SCWO process can mineralize organic materials by reacting them with an oxidant (often O_2 in air) in water above its critical temperature and pressure [T_c = 705°F (374°C), P_c = 218 atm (3,200 psi)]. Organic carbon is converted to CO_2, hydrogen is converted to water, and N, S, and P heteroatoms are converted to N_2, sulfate, and phosphate, respectively. The elementary chemical reactions that occur during SCWO are analogous to those that occur during combustion. However, the lower temperature, higher pressure, and abundance of water in SCWO alter some of the reaction pathways from those expected in combustion.

When a mixture is above its thermodynamic critical temperature, it can exist only as a single fluid phase. Coexisting liquid and gas phases cannot form. Thus, SCWO allows the organic material, the oxidant, and water to exist in a single fluid phase under reaction conditions. This absence of fluid phase boundaries prevents interphase transport processes from limiting reaction rates. Consequently, SCWO reactions are rapid, and complete mineralization can often be achieved in seconds.

Since the oxidation reaction is exothermic, the temperature of the reactor effluent [e.g., 1,112°F (600°C)] typically exceeds that of the reactor feed. This thermal energy from the reaction can be captured in a well-engineered process through heat integration and then used again in the process. In short, SCWO provides an option for the rapid and nearly complete oxidative destruction of organic material. The process allows for analysis of the effluent before release so that one can be assured that desired destruction efficiencies are being achieved.

The SCWO environment can be corrosive, especially if a region exists in the process where the aqueous feed stream is in the dense, near-critical state [around 572-662°F (300-350°C)]. The presence of halogens also increases corrosion rates. There have been different approaches to handling corrosion in SCWO systems, including preventing corrosive species from reaching reactor surfaces, altering the feed stream composition or processing conditions to reduce corrosion, or managing corrosion by using sacrificial reactor liners made from materials such as titanium.

Though liquid water is a good medium for dissolving salts, supercritical water is not. It has a much lower density and far fewer hydrogen bonds, so it is not a favorable medium for ion formation. Indeed, ions in supercritical water typically exist as ion pairs. Thus, any salts present in the feed stream and any salts that form during SCWO may precipitate under SCWO operating conditions. This can lead to salts accumulating in, and plugging, the SCWO reactor. Different methods have been examined for dealing with salt precipitation or accumulation in SCWO reactors. These methods include avoiding precipitation by operating at high densities (very high supercritical pressures), allowing precipitation but avoiding accumulation (e.g., by keeping the salts mov-

[4] Annual Status Report on the Disposal of Chemical Weapons and Materiel for Fiscal Year 2007, www.peoacwa.army.mil/wp-content/uploads/annual_status_report_disposal_of_chemical_weapons_materiel_fy07.pdf.

[5] Battelle Calculation Continuation Sheet, Concentration of PCBs in one EBH batch, no date.

ing through the reactor), and allowing both precipitation and accumulation but periodically removing the salt deposits from the reactor walls (e.g., by brushing or scraping).

Supercritical Water Oxidation at BGCAPP

BGCAPP has three identical vertically-oriented SCWO tubular reactors that will operate in continuous flow mode at about 1,150°F (621°C) for GB hydrolysate and 1,175°F (635°C) for VX hydrolysate, and 235 atm (3,454 psi). Each reactor is 10 ft high and 7-5/8 in. in diameter. The reactor residence time is about 10 sec. Hydrolysate, fuel (isopropanol), and air enter the top of the reactor and effluent exits from the bottom, where quench water is added (see Figure 2-1). Each reactor can process 1,000 lb/hr of blended hydrolysate feed.

Both corrosion and salt management are relevant concerns for the SCWO process at BGCAPP. Phosphate and fluoride ions, which are known corrosive agents, will be present in the feed to the SCWO reactor. BGCAPP will deal with corrosion by employing a grade 2 titanium reactor liner that will be periodically replaced as it corrodes. Replacement every 300-400 hours is anticipated, depending on the agent hydrolysate being treated. Salts will be managed by adding sulfur and chloride to the feed such that a mixture of NaCl and Na_2SO_4 will coprecipitate from solution in the SCWO reactor. This salt mixture is a liquid under a known range of conditions, so it will flow down the vertical reactor to the quench zone at the bottom and should not cause deposits or plugging in the upper reactive reaches of the SCWO reactor.

As discussed above, BGCAPP will use SCWO to treat blended agent (GB or VX) and energetics hydrolysates. Previous pilot-scale work verified that SCWO can achieve high destruction efficiencies for actual hydrolysate produced from both nerve agents and from energetic materials as well as for blended hydrolysate (BPBGT, 2014). Additionally, FOAK testing showed that the reactors to be used at BGCAPP can operate successfully for extended periods of time when the feed stream is simulated hydrolysate. FOAK testing provided 756 hr (not continuous) of reactor operation with simulated hydrolysates (BPBG, 2013). FOAK testing showed that the total organic carbon content in the effluent was <5 ppm and that the salts management strategy was effective. Moreover, the FOAK testing showed that replacement of the thermowells would need to be more frequent (about every 100 hr) than replacement of the SCWO reactor liner (about every 300-400 hours of operation) (BPBGT, 2014).

Finding 2-2. Only hydrolysate simulants have been processed through the full-scale SCWO unit during FOAK testing. Actual hydrolysate, however, has been tested at the pilot scale.

Finding 2-3. SCWO has been demonstrated to be a robust method for treating a wide range of organic wastes, including both actual and simulated hydrolysates. Lack of knowledge of the precise composition of the hydrolysates and their variability does not diminish confidence in the ability of SCWO to destroy the organic carbon present therein.

DESCRIPTION OF THE BGCAPP WATER RECOVERY SYSTEM

A reverse osmosis (RO) system is to be used to desalinate a blend of SCWO effluent, cooling tower blowdown water, and steam boiler blowdown water.[6] Water softeners will be used to remove calcium from the cooling tower and steam blowdown water, to avoid fouling of the RO units. According to BPBG, 2007, ion exchange beds will be used to soften this stream prior to blending it with the SCWO effluent. The RO system is intended to produce a permeate stream and a reject stream. The permeate will be recycled as quench water for the SCWO reactors. The spent softener regenerant[7] will be combined with RO reject water for final disposal.

The WRS system was designed to do the following:

- Operate with an efficiency of 70 percent water recovery with a maximum of 500 mg/L total dissolved solids in the permeate and
- Ensure one and a half day's storage of RO permeate to permit SCWO operation in case the WRS is temporarily not operating.[8]

To accomplish these operations, the WRS will have the following equipment:

- Three SCWO effluent storage tanks, where the effluent will be analyzed to ensure that the total organic carbon (TOC) concentration is less than 10 ppm;
- A conventional pretreatment system consisting of coagulation, media filtration, and antiscalant addition;
- Three spiral wound RO units (two operational, one spare); and
- Storage tanks to hold RO permeate to periodically clean the RO membranes and to provide SCWO reactor quench water.

The pretreatment portion of the WRS will remove suspended solids, while the RO system will reduce the total dissolved solids. Figure 2-2 shows the flow of material from hydrolysis, through the SCWO process (shown in

[6] Blowdown water is water that is drained from cooling equipment or boilers to remove minerals that accumulate over time. By definition, they tend to concentrate calcium and other dissolved impurities.

[7] The regenerant is the waste solution resulting when a high concentration solution of NaCl is used to renew the hardness removal capacity of the ion exchange resin. It typically contains high concentrations of NaCl, calcium salts, and magnesium salts.

[8] NRC Request Update, NRC#08, received via e-mail on February 16, 2015.

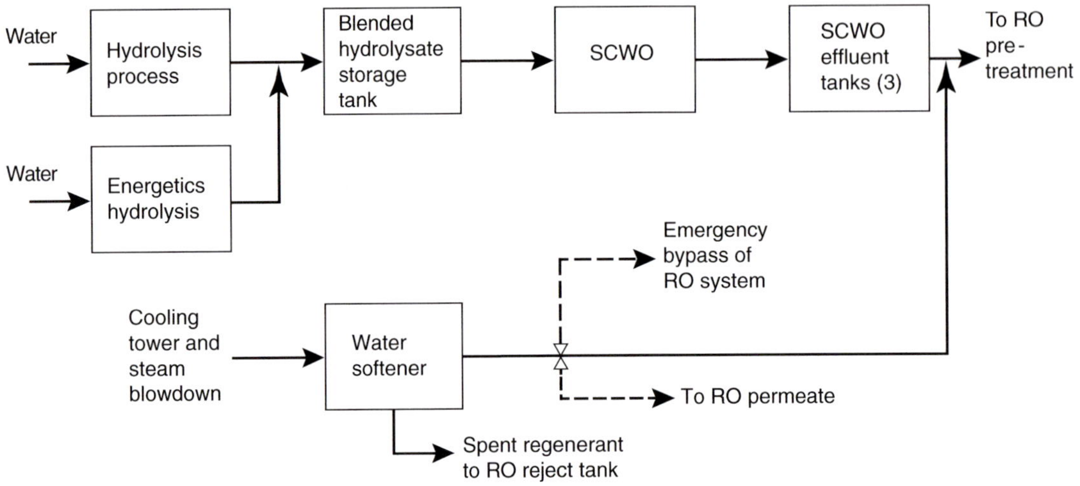

FIGURE 2-2 Flow of material from hydrolysis, through SCWO, and until the pretreatment step in the WRS. SOURCE: NRC, 2012.

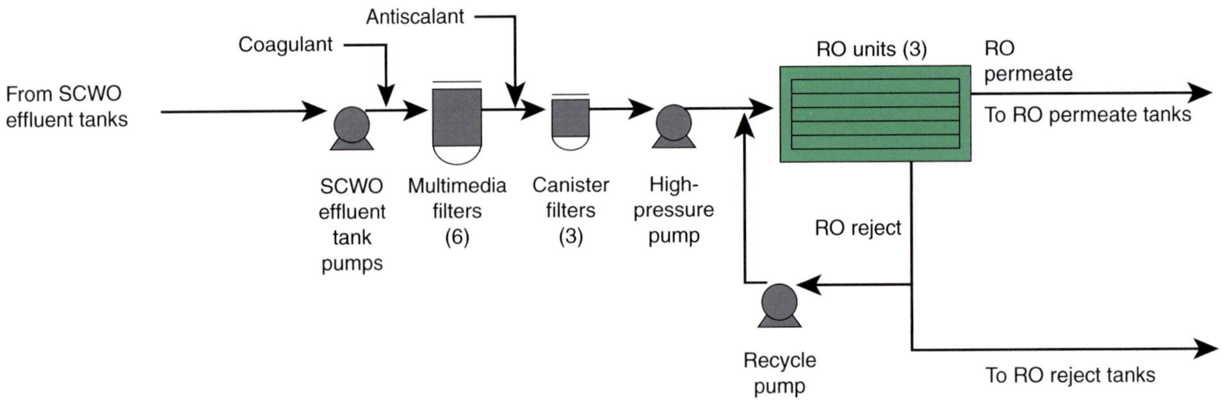

FIGURE 2-3 Process flow diagram for the WRS, including the pretreatment and RO system. SOURCE: NRC, 2012.

Figure 2-1), up to the pretreatment step in the WRS. It also indicates where the cooling tower and steam blowdown is blended with the SCWO effluent. Figure 2-3 shows the flow of material through the WRS.

REFERENCES

Bonnett, P.C., and B. Elmasri. 2002. *Base Hydrolysis Process for the Destruction of Energetics Material.* AD-E402 946. Picatinny Arsenal, N.J.: U.S. Army Armament Research, Development and Engineering Center.

BPBG (Becthel Parsons Blue Grass). 2007. SCWO Building Water Recovery—R.O. Unit Process Flow Diagram, Rev. 6, Sheet 1/2, October 2. Richmond, Ky.: Blue Grass Chemical Agent Destruction Pilot Plant Project.

BPBG. 2013. *Test Report for Supercritical Water Oxidation (SCWO) First-of-a-Kind (FOAK) Test (Preliminary Draft).* SDN-24915-00-GQY-GGEN-00094. Richmond, Ky.

BPBGT (Becthel Parsons Blue Grass Team). 2014. *Characterization of Supercritical Water Oxidation (SCWO) Operating Requirements, Rev. 0.* 24915-00-30R-SCWO-00001. Richmond, Ky.

Dejarme, L., and G.D. Lecakes. 2008. *Bench-Scale Evaluation of VX Hydrolysis, TRRP #11, Test Report, Rev. 2.* Aberdeen, Md.: Battelle Eastern Science and Technology Center.

Malloy IV, T.A., L. Dejarme, C. Fricker, J. Guinan, G.D. Lecakes, and A. Shaffer. 2007. *Bench-Scale Evaluation of GB Hydrolysis, TRRP #02a Phase II, Test Report, Rev. 0.* Aberdeen Proving Ground, Md.: Program Manager for Assembled Chemical Weapons Alternatives.

NRC (National Research Council). 2012. *Letter Report: The Blue Grass Chemical Agent Destruction Pilot Plant's Water Recovery System.* Washington, D.C.: The National Academies Press.

NRC. 2013. *Assessment of Supercritical Water Oxidation System Testing for the Blue Grass Chemical Agent Destruction Pilot Plant.* Washington, D.C.: The National Academies Press.

3

Stakeholder Interests and Issues

In this chapter the committee provides an overview of the current public involvement process and how past experience with decision making and public involvement has shaped stakeholders' views about how to address the potential for offsite shipments of hydrolysate. A variety of stakeholders are interested in and affected by activities at the Blue Grass Chemical Agent Destruction Pilot Plant (BGCAPP) facility. The discussion in this chapter focuses on the interests and involvement of local members of the public, relying primarily on the Citizens' Advisory Commission (CAC) and the Chemical Destruction Citizens' Advisory Board (CDCAB), an independent subcommittee of the CAC, which serve as the institutional representatives of the local and state populations.[1] Public involvement has been highlighted as a critical component of the Assembled Chemical Weapons Alternatives (ACWA) program since its inception.[2] Indeed, the Program Executive Office (PEO) ACWA "attributes its success in identifying safe and effective alternatives for chemical weapons destruction to its commitment to meaningful stakeholder input and involvement."[3]

The chapter concludes with a series of findings and recommendations. Appendix A tells the history of public involvement in the ACWA program.

CURRENT PUBLIC INVOLVEMENT PROCESS

The public involvement process at BGCAPP is well established and includes a variety of opportunities for stakeholders to obtain consistent information about and become involved in project activities and decisions. The outreach effort is described as a team effort among governmental and contractor staff from PEO ACWA, BGCAPP, the Blue Grass Army Depot (BGAD) and the Blue Grass Chemical Activity.[4] PEO ACWA is responsible for providing information about operations and programs to the broader public by maintaining an informational website, for responding to media enquiries and, since 2010, for periodically posting on Twitter and Facebook.[5] Additionally, PEO ACWA provides a quarterly briefing that is posted on the PEO ACWA website.[6] The local outreach office, staffed by three full-time and one part-time contractor staff, is often the first line of engagement for local residents and is responsible for a variety of activities. The staff distributes monthly electronic newsletters to a mailing list estimated at approximately 1,800 subscribers as well as to community leaders and to specialized mailing lists for news releases. Additional messages are also distributed from the BGCAPP project manager. The local outreach office contractor staff pass media enquiries to PEO ACWA headquarters, but they field enquiries from the public; support and facilitate CAC and CDCAB meetings; produce information products under government direction; participate in local activities, including staffing information booths at local events; operate a speakers bureau; maintain an active educational program for classes ranging from kindergarten through college; and arrange site tours.[7]

The CAC, established by Kentucky statute in 1994, comprises nine members appointed by the state governor. The

[1] Stakeholders include personnel from the PEO ACWA, BGCAPP, Blue Grass Chemical Activity and Blue Grass Army Depot, the Kentucky Department for Environmental Protection; members of the CAC/CDCAB, other local members of the public, and members of the Kentucky Environmental Foundation and Chemical Weapons Working Group.

[2] The program was first established as the Assembled Chemical Weapons Assessment program. It subsequently became the Program Manager for Assembled Chemical Weapons Alternatives and, finally, the Program Executive Office for Assembled Chemical Weapons Alternatives.

[3] https://www.peoacwa.army.mil/media-toolkit/facts-pages/peo-acwa-legislation/.

[4] April 7, 2015, conference call with Miguel Monteverde, public affairs specialist, PEO ACWA, and Sarah Parke, manager, Blue Grass Chemical Stockpile Outreach Office; Judith Bradbury and Seth Tuler, committee members; Todd Kimmell, committee chair; and Jim Myska, study director, NRC.

[5] Blue Grass Chemical Agent-Destruction Pilot Plant (BGCAPP), http://www.peoacwa.army.mil/bgcapp/.

[6] Ibid.

[7] April 7 conference call, op. cit.

CDCAB, formed in 2003, functions as a subcommittee of the CAC to provide broader stakeholder involvement from a range of local organizations. These include government, civic, medical, emergency management, and university and school representatives. Of these CDCAB members, 24 are voting members and 7 are representatives of the agencies being advised by the body.[8] The CAC/CDCAB members meet jointly four times a year.[9]

COMMITTEE APPROACH TO GATHERING INFORMATION

The committee gathered information about the interests and viewpoints of local community members from several sources:

- Open discussion at the committee's January 2015 meeting in Kentucky;
- A public meeting, advertised in the local media and on the BGCAPP outreach office website, that was scheduled during the January 2015 committee meeting;
- Establishment of a dedicated e-mail address to which comments could be submitted. The address was printed on business cards, which were publicized in the local media and on the PEO ACWA website, distributed at the public meeting, and made available at the chemical stockpile outreach office (see Appendix B);[10] and
- Follow-up telephone interviews with the CAC chair, a CDCAB co-chair, and PEO ACWA and BGCAPP outreach staff.

In addition, the CAC/CDCAB had been briefed by PEO ACWA and was given an opportunity to comment on this committee's statement of task for the BGCAPP hydrolysate study prior to its issuance. The committee held its first meeting for this report in Richmond, Kentucky, close to BGCAPP, in January 2015 and invited CAC/CDCAB members to join it for the presentations and open discussion. Two members of the CAC/CDCAB, including the CAC chair and one co-chair of the CDCAB,[11] attended the two days of open meetings and the concurrent public meeting. As part of the formal discussions and presentations at the committee meeting, the CAC/CDCAB provided written and verbal suggestions and expectations for the conduct of the study, including criteria that they believe should guide a decision to initiate offsite shipment of the hydrolysate should it be necessary. The committee and public meetings also provided an opportunity for committee members to introduce themselves and interact informally with the local representatives, as well as providing an overview of the study, responding to questions, and further emphasizing the importance of community input.

The committee also attempted to gather information from the broader public. The committee's site visit and public meeting were advertised in the local media and on the PEO ACWA website; however, only one member of the public attended and no input was received other than from the CAC/CDCAB.[12] Both the CAC/CDCAB and outreach staff from the site and PEO ACWA attributed the lack of input to residents' trust in the CAC/CDCAB to protect their interests and reported little evidence of recent community discussion about possible offsite hydrolysate shipment or the risks of continued storage of the chemical weapons. The CDCAB co-chair emphasized that, if they had any concerns, the public was quick to contact CAC/CDCAB members and stated that "We hear from them . . . don't think it's because they don't care about it."[13]

PAST EXPERIENCE AS IT AFFECTS THE CONTEXT FOR ASSESSING CONTINGENCY OPTIONS

CAC/CDCAB members' past experiences shape the way they currently approach consideration of any contingency option for the disposal of hydrolysate (see Appendix A for a more detailed discussion). These experiences include (a) the ACWA Dialogue process and (b) the subsequent development of public involvement at BGCAPP. In combination, these experiences have contributed to

- The emergence of "critical trust" after the severe erosion of trust between community members and Army during early program efforts to incinerate chemical weapons;[14]
- An expectation by community members that they will continue to play a meaningful role in decision making about facility design, monitoring, and performance; and
- Continued, principled opposition to offsite shipment of the hydrolysate.

[8] Blue Grass Chemical Agent-Destruction Pilot Plant (BGCAPP), http://www.peoacwa.army.mil/bgcapp/.

[9] According to the outreach staff, the organizations are typically referred to jointly as the CAC/CDCAB. All CAC members are CDCAB members, but not all CDCAB members are also CAC members.

[10] NRC Seeks Public Comment on Hydrolysate Transport, http://www.richmondregister.com/news/nrc-seeks-public-comment-on-hydrolysate-transport/article_ead95fb8-a1cd-11e4-a983-bf9e2a947b10.html.

[11] The position of second co-chair was in transition at the time of the committee's site visit.

[12] The BGCAPP chemical stockpile outreach office manager reported that the cards providing the e-mail address for comments were publicized on the PEO ACWA website and Facebook, made available on the reception desk at the outreach office, and included in various presentations given by the office staff.

[13] Discussion during the committee meeting in Richmond, Kentucky, January 2015.

[14] Poortinga and Pidgeon (2003) call "critical trust" "a practical form of reliance on a person or institution combined with some healthy skepticism" (p. 971). It can serve important social functions, such as ensuring oversight and vigilance.

The ACWA Dialogue Process

Beginning in the 1980s and, more notably during the 1990s, the chemical weapons destruction program was stalled by increasingly vocal and active opposition to incineration of the stockpile by residents living near some of the chemical weapons storage sites, in particular, the Pueblo Chemical Depot and BGAD. Prior to establishment of the ACWA program, residents believed that the Army discounted their questions and concerns about the safety of incineration and the safety of their families and communities. Pressure from Congress, spearheaded by the Kentucky Environmental Foundation and the Chemical Weapons Working Group (CWWG), was among the factors that led to the establishment of ACWA.[15]

As discussed in Appendix A, ACWA's subsequent five-year-long Dialogue on Assembled Chemical Weapons (the ACWA Dialogue), initiated in 1997 and facilitated by personnel from the Keystone Policy Center, included input into the evaluation of alternative technologies for which the Army solicited proposals.[16] It is important to note that the CAC chair and the CDCAB co-chair were involved in the original ACWA Dialogue group. The CAC chair described the ACWA Dialogue as "a remarkable accomplishment" and "a wonderful model of decision making. The key is to get everyone at the table, involve them in planning from the very beginning and not after a decision is made."[17] Interviews that the committee held with the CAC chair and CDCAB co-chair demonstrated that the experience continues to affect their interpretation and expectation of *meaningful* public involvement at BGCAPP.

A number of aspects of the ACWA Dialogue policy process were notable from the perspective of CAC/CDCAB members:

- Inclusion of both technical staff and local, lay representatives from the chemical weapons sites;
- Inclusion of both technical and social criteria for evaluating the technologies;
- The unprecedented extent of involvement afforded to local representatives, including establishment of a Citizens' Advisory Technical Team that observed the scoring and weighting of evaluation criteria for the technologies;
- The community's "ownership" of the technology as a result of its input and close involvement; and
- The beginning of a transformation from antagonism to an attitude of understanding and trust.

Development of Meaningful Public Involvement

In a telephone interview with the committee, the CAC chair and the CDCAB co-chair described the building of trust, transparency, and collaboration between PEO ACWA and the community around BGCAPP as a long-term process. The process began with the establishment of the ACWA program and the ACWA Dialogue and continued with the mandate for public involvement and the appointment of an ACWA program officer who understood the value of, and the need for, transparency and meaningful public involvement.[18] The CAC also recognized the need for broader representation of the community, leading to establishment of the CDCAB as an entity that would ensure input was sought from, and communication was maintained with, people representing the many groups and interests that make up the local community. Shortly afterwards, the CAC established working groups to study particular issues in detail, including a Secondary Waste Working Group, which has been actively engaged in the discussion of waste issues for over a decade and reports back to the joint meetings of the CAC/CDCAB. The CAC thus used the leverage provided by the PEO ACWA statutory directive for public involvement to push for what members envisioned as *meaningful* public involvement.

Outreach personnel from BGCAPP and PEO ACWA headquarters, as well as the CAC chair and the CDCAB co-chair, describe the current relationships between PEO ACWA and the community as very positive. For example, in his opening statement at the January meeting, the CAC chair emphasized the importance of PEO ACWA's openness and expressed appreciation for the trust and relationships that PEO ACWA had built with the community that had turned around community residents' original opposition and lack of trust. The CDCAB co-chair subsequently expanded on the evolution of relationships and opportunity for public involvement that had occurred, and stated as follows:

> Historically, when decisions were brought to the community, there was opposition and argument because there was no discussion with the community beforehand. Now, topics are brought to us well before, for example, decisions about design changes are brought to us even before the structure of the process or a decision is in place. An example is that we got the [National Research Council] statement of task to review—we learned about what was being proposed by ACWA and have been heavily involved in the process since. A remarkable change from coming to town and announcing that we are going to incinerate. So now, even though we may not like a design change, we are more likely to accept

---

[15] The CWWG includes members of the public living near chemical weapons storage sites across the U.S. and of national and international organizations. It is now less active than formerly because there are fewer sites where weapons remain. The CDCAB co-chair was a founding member of these organizations and currently identifies himself as Chemical Weapons Program Director of the Kentucky Environmental Foundation and national spokesperson for the (now smaller) CWWG.

[16] Formerly named the Keystone Center, this Colorado organization is dedicated to facilitating resolution of national policy issues.

[17] March 25, 2015, conference call with Doug Hindman, Kentucky CAC chair, and Craig Williams, Kentucky CDCAB co-chair; Judith Bradbury and Seth Tuler, committee members; Todd Kimmell, committee chair; and Jim Myska, study director, NRC.

[18] Ibid.

> it because we recognize it is for the safety of the community, we accept that it is the best way to go. For example, 15 years ago, no one [community stakeholders] would have considered as acceptable the [explosive destruction technology] or the thermal oxidizer [for the metal parts treater]. If we had not had the relationships, transparency, trust building between the Army and the community, that would have been a non-starter. But now, we can come together to agree about things that historically would have been distasteful or disagreeable and look at things in a broader sense.[19]

KEY CONCERNS EXPRESSED BY THE CAC/CDCAB

Several themes arose among the concerns expressed by the CAC chair and the CDCAB co-chair during the January 2015 meeting and in a subsequent telephone interview. These included the CAC/CDCAB's continuing opposition to offsite shipment of the hydrolysate, along with their very reluctant and conditional acceptance of such an outcome as a last-resort contingency plan; their concerns about the impact of offsite hydrolysate shipment on receiving communities; and their perspective on the criteria for making a decision on whether offsite shipment of the hydrolysate would be warranted. Additionally, the CAC chair and the CDCAB co-chair outlined their expectations for this committee's study and, in particular, their expectation that the CAC/CDCAB would play an active role in decision making by participating in the development of criteria for determining whether and how future offsite shipments of hydrolysate occur and by participating in discussions of emerging operational issues (see written statement in Appendix B).

Opposition to Offsite Shipment

The CAC/CDCAB's opposition to offsite shipment of hydrolysate has a long and well-documented history.[20] For example, a statement provided to the National Research Council (NRC) and developed jointly by the Colorado and Kentucky CACs in 2008, lists specific concerns about offsite shipment of hydrolysate:

- Increased risks associated with transportation,
- Opposition from receiving communities,
- Negative economic impact on the local community,
- Probable cost increases and schedule slippage,
- Inaccurate and inflated cost savings attributed to offsite shipment,
- Political opposition,
- Possible litigation,
- Risk to the BGCAPP permit caused by eliminating onsite secondary treatment,
- Violation of environmental justice principles, and
- Elimination of potential legacy use of onsite treatment facilities—for example, those at BGAD (NRC, 2008).

In his statement to the committee in January 2015, the CAC chair confirmed the CAC/CDCAB's continuing opposition to offsite shipment of hydrolysate. However, he also emphasized that the chemical munitions stored at BGAD must be destroyed as soon and as safely as possible. Furthermore, he expressed his concern that the supercritical water oxidation (SCWO) treatment process might fail to perform as expected. He opined that members of the CAC/CDCAB recognize there are bound to be issues with a pilot plant such as BGCAPP, and failure of the supercritical water oxidation or other key processes could therefore pose a dilemma for them.[21]

Decision Criteria

The CAC/CDCAB presented to this committee a set of criteria for offsite shipment of hydrolysate, should such a move be deemed necessary. Their bottom line is that offsite shipment would be considered acceptable if it were the only alternative and "only for that portion of the neutralization output that cannot be processed through planned secondary treatment and/or stored onsite until mitigation is achieved."[22] Conditions that might necessitate offsite shipment were suggested:

- If secondary treatment of hydrolysate is unable to keep up with neutralization output,
- All feasible mitigation measures have been evaluated, and
- Mitigation is impossible or would require that plant operations be suspended for an extensive period of time ("extensive" should be defined within the study).[23]

In addition, the CAC/CDCAB recommended that the committee consider alternative onsite treatment options should onsite downstream treatments fail. One such option is modifying the explosive destruction technology that will be used to destroy the mustard munitions at BGAD to process excess hydrolysate. Their recommendation for alternative onsite treatment is addressed in Chapter 1.

CAC/CDCAB Expectations for Public Involvement

The CAC/CDCAB's written statement and interview with the committee reflected their very active involvement

[19] Ibid.

[20] Their opposition to the offsite shipment was reinforced by the CDCAB co-chair's link to the Kentucky Environmental Foundation and the national CWWG. See Appendix A for more detail.

[21] In the past, the CAC/CDCAB agreed to offsite shipment as part of Operation Swift Solution, but at the same time stated that their agreement and the program should not be viewed as a precedent.

[22] CAC/CDCAB written statement to the committee (see Appendix B).

[23] Ibid.

and positive experiences in the ACWA Dialogue process and their subsequent participation in BGCAPP activities and decisions, as discussed briefly in this chapter and in greater detail in Appendix A. In their statements, they made clear their expectation that they would play an active role alongside PEO ACWA, site staff at BGCAPP and BGAD, and Kentucky regulators in (1) developing protocols that identify the circumstances under which offsite shipment of hydrolysate would occur and (2) developing a means for information sharing and providing input on issues that bear on any decision regarding offsite shipment, such as the level of hydrolysate storage that might trigger shipment. They also noted the need to address the tension between proactive planning related to regulatory requirements on the one hand and facilitating the process for shipping hydrolysate offsite on the other—in other words, the concern that advanced planning would make it easier to decide to ship hydrolysate offsite.

The CAC chair and the CDCAB co-chair described the current process for involvement in decisions as a transparent "back-and-forth relationship" and that "if there is an issue that needs to be raised by us or them we work through it incrementally and work towards agreement about how to proceed. We have done this all the time and it has worked each time." While recognizing that their legislated role is consultative and that they have "no authority to do anything," they nevertheless noted that "we carry some weight because of how we operate. . . . If something is repulsive to the community, we have a history of 'making it known.' "[24]

In discussing the CAC/CBCAB perspective on how BGCAPP should make decisions about shipping hydrolysate offsite, they said "this is something we want to work out with [BGCAPP], we want to be part of the process" and "the whole purpose of this exercise is to do that with them." They acknowledged that many details are involved that require additional consideration. For example, the question of what percentage of total hydrolysate storage capacity being reached would serve as a threshold for offsite shipment could not be answered at this time. Rather, "we need to address that question legitimately through a process, we want to be involved in that process—the CAC and CDCAB."[25]

When asked how they envisaged their involvement in developing criteria or in making decisions that could lead to offsite shipment, the CDCAB co-chair stated that "what the NRC report comes up with about decision points will be very interesting for us, and a starting point for many discussions here," and provided the following, detailed description of the process they envisaged:

> ACWA is already discussing this with us and we assume it will continue. The way we see it is the NRC report is a tool that ACWA will use in the context of what they need to consider on this aspect of things. They and we will read the report, we will sit down—the Secondary Waste Working Group will deal with it first and will come to some draft proposal that will include criteria points associated with a contingency plan and then bring it to the full body of the CAC to create a recommendation to ACWA. Hopefully by that time, recommendations will emerge that are acceptable to the regulators [Kentucky Department for Environmental Protection], CAC, the Secondary Waste Working Group, Bechtel, Army—there will be a series of meetings so that ultimately we will get a recommendation about how to proceed. That is the process that will be used. Once the NRC report comes out, I am sure there will be a series of meetings to walk through the options and weigh against criteria, and also the considerations embedded in draft recommendations. Bechtel Parsons and ACWA, in the normal course of events, will participate and present their insights and recommendations at our meetings, so they will know where we are moving and what we are discussing. In this way, we will get everyone together as early as possible and there are no surprises. . . . We said it earlier—no surprises. You don't surprise us, we don't surprise you. No surprises on both sides, that is the mantra of the chemical weapons program in Kentucky.[26]

Concerns about the Impacts of Offsite Shipment on Receiving Communities

Concerns about meaningful public involvement expressed by the CAC/CDCAB extend to potential receiving communities as well, in the event that offsite shipment of hydrolysate is deemed necessary. In common with the CWWG and the Kentucky Environmental Foundation, the CAC/CDCAB firmly believe in environmental justice principles, which they interpret as requiring that a community take care of its own wastes and not impose them on other, perhaps more vulnerable, communities. Consequently, they strongly believe that sites scheduled to receive hydrolysate shipped offsite need to be engaged in planning for such shipment.

Consistent with this viewpoint, the CAC chair and the CDCAB co-chair expressed concern that the committee's statement of task was too narrow and did not extend to consideration of the location and concerns of communities that could potentially receive any hydrolysate shipments. In discussing their expectations for the NRC study during the committee's January 2015 meeting, they emphasized the importance of identifying the location of any potential site or community that would receive the waste, as well as information on the technology that would be used, as part of planning and decision making for offsite shipment of hydrolysate in the event that such disposal is deemed necessary. This would include the schedule impact from potential regulatory, legal, and political opposition in the receiving communities. They emphasized in the March 25, 2015, telephone conversation their concern that the residents of these communities need to know in advance about potential shipments and have an opportunity to take part in decisions that would affect them; the co-chair of the CDCAB said "it

[24] March 25, 2015, conference call, op. cit.
[25] Ibid.
[26] Ibid.

is not legitimate to spring them [hydrolysate shipments] on a community that has not been involved in the decision—it is not an acceptable or ethical approach when dealing with these kinds of materials."[27] They recommended considering past experiences with offsite hydrolysate shipment, including those that were quite controversial. The CDCAB co-chair warned also that trust and acceptance are not guaranteed and that, if hydrolysate shipments are needed, PEO ACWA and all parties involved would have to work out a process for involving the receiving communities to avoid confrontation with the communities and with the CWWG, as had previously happened with hydrolysate shipments from the Newport Chemical Agent Disposal Facility in Indiana. Additionally, he compared the shipments from Indiana with past shipments conducted from BGCAPP. As part of Operation Swift Solution at BGCAPP, the "community was engaged with our assistance" and shipments were accomplished "with no protests, no lawsuits, no opposition, no politics . . . in part because it was a smaller amount, but more importantly because they were part of a process before the decision was made."[28]

SUMMARY

As demonstrated in discussions and by their statement provided to the committee in January 2015, the CAC/CDCAB recommended that emphasis be placed on the contingent nature of this study. The CAC/CDCAB's role in decision making was an important theme. They continue to oppose offsite shipment of hydrolysate, and they regard the agreement on technologies planned for use at BGCAPP and reached as a result of the ACWA Dialogue process as a commitment to the community. They recognize, however, that operational issues are bound to arise in a pilot plant, and that it may not be possible to fulfill PEO ACWA's commitment to onsite treatment. From their perspective, PEO ACWA's commitment is one that can be abandoned only as a final resort, and as much as possible as a temporary measure, when no other onsite treatment options exist (see Appendix B, CAC/CDCAB written statement).

The ACWA Dialogue experience was pivotal for local participants around BGCAPP. Their experience in that process has since been reinforced by their experience of meaningful involvement in decisions regarding BGCAPP and the building of transparency and trust between PEO ACWA and the community. Indeed, based on these experiences, CAC/CDCAB members hold high expectations for the scope and conduct of this study and for their role in decisions concerning potential offsite hydrolysate shipment. Their expectations include, in particular, opportunities for CAC/CDCAB members to provide input into analyses, evaluations, and the development of criteria identifying under what conditions offsite hydrolysate shipment would occur. Furthermore, they expect PEO ACWA will continue to provide regular updates on the status of operations as they evolve, to identify the technology and the location of receiving communities, and to consider previous positions on hydrolysate shipments of such communities and the CAC/CDCAB, including proactive information sharing and meaningful engagement.

FINDINGS AND RECOMMENDATIONS

Finding 3-1. The local and regional public, as represented by the CAC/CDCAB, agree on the need to destroy the chemical munitions stockpile at BGAD as soon as possible and in reducing risk and ensuring the safety of workers and the community.

Finding 3-2. The CAC/CDCAB continue to oppose offsite transport of the hydrolysate and to regard the agreement reached in the Assembled Chemical Weapons Alternatives Dialogue process for onsite treatment of hydrolysate as a commitment to the community around BGCAPP.

Finding 3-3. The CAC/CDCAB recognize that offsite shipment of the hydrolysate may be warranted. But, they qualify this recognition by insisting that any offsite shipment of hydrolysate should be viewed as much as possible as a temporary measure to be initiated only as a final resort.

Finding 3-4. The Program Executive Office for Assembled Chemical Weapons Alternatives, BGCAPP, the CAC, the CDCAB, and the community have invested considerable effort in building a solid foundation of trust and transparency and a workable institutionalized structure for meaningful stakeholder involvement. However, continued trust and acceptance must be nurtured.

Finding 3-5. The Assembled Chemical Weapons Alternatives Dialogue process and subsequent public involvement opportunities have established high expectations about how and when the CAC/CDCAB should be involved.

Finding 3-6. Should offsite shipment of hydrolysate be deemed necessary, public concerns about possible impacts on the receiving communities could stall such shipments.

Recommendation 3-1. In collaboration with the CAC/CDCAB, the Program Executive Office for Assembled Chemical Weapons Alternatives should institutionalize a transparent consultation process that builds on the existing foundation and working group structure to ensure meaningful stakeholder input into analyses, evaluations, and decision criteria related to potential offsite shipment of hydrolysate

[27] Ibid.

[28] Operation Swift Solution involved the disposal of three deteriorating ton containers that held GB agent and related breakdown product. The GB in these containers was chemically neutralized in a special facility at BGAD and the hydrolysate was shipped offsite for final disposal, http://www.peoacwa.army.mil/bgcapp/about-bgcapp/operation-swift-solution/.

and that provides opportunities for engaging with communities that would receive hydrolysate.

Recommendation 3-2. To maintain the existing policy of "no surprises" and to help alleviate concerns about proactive planning, agreement about the process for consultation should be in place before the BGCAPP initiates advance regulatory and logistical planning for offsite shipment of hydrolysates.

REFERENCES

NRC (National Research Council). 2008. *Review of Secondary Waste Disposal Planning for the Blue Grass and Pueblo Chemical Agent Destruction Pilot Plants.* Washington, D.C.: The National Academies Press.

Poortinga, W., and N.F. Pidgeon. 2003. Exploring the dimensionality of trust in risk regulation, *Risk Analysis* 23: 961-972.

4

Regulatory Requirements for Offsite Hydrolysate Shipment and Treatment

INTRODUCTION

The primary mission of the Blue Grass Chemical Agent Destruction Pilot Plant (BGCAPP) is to safely destroy the Blue Grass Army Depot (BGAD) chemical stockpile, including meeting criteria for successfully treating the resulting hydrolysates in the supercritical water oxidation (SCWO) and water recovery system (WRS) processes. BGCAPP is currently operating under a Kentucky Department for Environmental Protection (KDEP) Resource Conservation and Recovery Act (RCRA) Research, Development and Demonstration (RD&D) permit for the destruction of GB munitions. Under the RD&D permit application Revision 5 submission, the destruction of the GB munitions will be used to demonstrate the effectiveness of the BGCAPP treatment processes. At the conclusion of the RD&D program, the intent is that the facility will have demonstrated that it is capable of operating at full capacity and the remaining VX munitions will be processed through BGCAPP pursuant to a modification to the existing BGAD RCRA (Part B) permit.

The RD&D permit provides for flexibility in implementing operational modifications to the process as part of systemization or pilot testing phases to address deficiencies that may develop or to improve efficiency or effectiveness. Data obtained during both ongoing equipment design and testing programs under the RD&D permit will be used to improve and validate BGCAPP processes, including the SCWO and WRS. The program schedule, as outlined in the RD&D permit application, is designed to allow sufficient time to identify and overcome minor problems that may develop. The RD&D permit will remain in effect for one year after GB munitions are first received, and may be renewed up to three times, with each renewal being for a period of up to one year. However, if unforeseen and insurmountable problems should arise with the SCWO or WRS, it could become necessary for BGCAPP to consider other options, including sending hydrolysate or SCWO effluent offsite in order to continue destruction of the GB and VX munitions currently stored at BGAD (see Chapters 6 and 7). Since such unforeseen and insurmountable problems could arise, it would be prudent to identify all regulatory requirements for offsite shipment and treatment of hydrolysates or SCWO effluent beforehand in coordination with KDEP and any other relevant regulatory bodies and with the Citizens' Advisory Commission (CAC) and the Chemical Destruction Community Advisory Board (CDCAB), the Program Executive Office (PEO) for Assembled Chemical Weapons Alternatives (ACWA), and other stakeholders so that the destruction mission is not unduly delayed.

RCRA PERMITTING

Regulatory Background

RCRA establishes a program for hazardous waste management from cradle to grave.[1] Treatment, storage, and disposal facilities (TSDFs) such as BGCAPP are required to obtain permits that establish specific operating conditions.

The federal program established by RCRA is administered by the U.S. Environmental Protection Agency (EPA). RCRA provides states with the option to seek EPA authorization to administer their own state-specific programs. Most states have this authorization. KDEP is presently authorized to administer most aspects of the RCRA program within the state of Kentucky.[2] KDEP's RCRA TSDF program is defined in the Kentucky Administrative Regulations (KAR), Title 401, Chapter 34, Standards for Owners and Operators of Hazardous Waste Storage, Treatment and Disposal Facilities (DEP, 2005).

[1] 42 U.S. Code §6901 et seq.: Code of Federal Regulations (CFR) at 40 CFR §§260 to 272.

[2] At this time, Kentucky has not received EPA state authorization for all RCRA regulations (e.g., the Hazardous and Solid Waste Amendments (HSWA) of 1984 or HSWA phased amendments). Therefore, federal standards and EPA oversight would apply to any unauthorized sections of the federal program.

Hazardous Waste Listings and Characteristics

While a state that administers its own RCRA program must maintain the standards in the federal RCRA program, it can also make its program more stringent and/or broader in scope. Kentucky's program is broader scope in that it lists specific chemical agents as acute hazardous wastes.[3] The Kentucky-specific listed hazardous wastes include chemical munitions containing the following agents: GB (isopropyl methyl phosphonofluoridate) and related compounds, VX (O-ethyl-S-(2-diisopropyl-aminoethyl)-methylphosphonothiolate) and related compounds, and HD (bis (2-chloroethyl) sulfide) and related compounds (waste codes N001, N002, and N003, respectively).[4]

In addition to establishing listed hazardous wastes, the RCRA program also establishes characteristic hazardous wastes, defined as wastes possessing characteristics such as ignitability, corrosivity, reactivity, and toxicity. A given hazardous waste can be a listed hazardous waste, a characteristic hazardous waste, or both (i.e., it is a listed waste that also possesses at least one RCRA-defined characteristic). Whether a waste is categorized as listed or characteristic is important because the residue of a listed hazardous waste is itself a listed waste (the RCRA derived-from rule), even if it no longer has the attributes of the original listed hazardous waste. This means that the hydrolysate and all downstream secondary wastes resulting from the treatment of listed agent wastes at BGCAPP will retain the Kentucky acutely toxic listed waste designation (i.e., N001 or N002).

It is anticipated that the energetics and agent hydrolysates will have a pH of 11-13 and therefore, in addition to being listed hazardous waste, hydrolysates will typically exhibit the RCRA corrosivity characteristic. These hydrolysates may also contain heavy metals and could therefore exhibit the RCRA toxicity characteristic.[5] The energetic portion of the munitions would meet the RCRA reactivity characteristic; however, it is anticipated that after hydrolysis, energetic hydrolysate will no longer be reactive (i.e., RCRA D003 waste).[6]

Unique to Kentucky, the specific listing of chemical munitions and their related compounds as acute hazardous wastes is required by Kentucky statute in addition to KDEP hazardous waste regulations.[7] RCRA provides a delisting process that is available for waste that the generator believes no longer meets the listing description.[8] However, in Kentucky, a change to the current RCRA acute hazardous waste listing for chemical munitions and their related compounds would require an amendment to the originating statute. It cannot be accomplished by a change to the KDEP regulation alone. The committee learned at its January 2015 BGCAPP meeting that the CAC/CDCAB's Secondary Waste Working Group, in coordination with BGCAPP, is currently contemplating a proposal for a legislative change that would not delist the chemical munition wastes but might add listed waste designations for specific treatment (e.g., derived) waste. These designations would apply to any treatment waste that could be shipped offsite for final treatment or destruction. The draft Proposed Waste Code Creation Table supplied by BGCAPP, dated December 18, 2014, shows 13 wastes that would be assigned new Kentucky waste codes, including agent hydrolysate and energetic hydrolysate. The presumption is that the BGCAPP hydrolysate and related secondary wastes proposed to be relisted under new waste codes will have met the 99.9999 percent destruction efficiency treatment standard set forth in the statute and TSDF regulations. The legislative change would relist these derived-from wastes such that they would no longer carry the original acutely toxic hazard code. They would, however, still carry a specific Kentucky hazardous waste code, and would still need to be managed as hazardous waste, just not acutely toxic hazardous wastes. It should be noted that SCWO effluent, which may need offsite shipment if the WRS underperforms or fails to perform, is not included in the current Proposed Waste Code Creation Table. At this time BGCAPP is not anticipating requesting that secondary wastes that meet the 99.9999 percent destruction efficiency treatment standard be delisted altogether.

Finding 4-1. Without a modification to the Kentucky statute that requires chemical munitions and their related compounds to be listed as acutely toxic hazardous wastes under the Kentucky RCRA program, and subsequent modification of the Kentucky Department for Environmental Protection regulations, all wastes derived from the treatment of chemical munitions at BGCAPP will retain the acutely toxic hazardous waste designation and will require additional handling and treatment requirements within Kentucky.

Recommendation 4-1. BGCAPP, in coordination with ACWA, should continue to support the CAC/CDCAB Secondary Waste Working Group in pursuing the legislative amendment and subsequent modification of Kentucky Department for Environmental Protection regulations such that all derived wastes can be stored, handled, and treated, both onsite and, as necessary, through offsite shipment, without the more burdensome requirements associated with acutely toxic hazardous wastes. In particular, SCWO effluent should be added to the list of wastes to be relisted with a new waste code.

---

[3] Acute hazardous wastes are wastes that contain chemicals so dangerous that they could pose a threat to human health and the environment even when properly managed. Possession of a much smaller quantity of acute hazardous wastes than other hazardous wastes will subject the generator to stricter onsite management standards than normal.

[4] See 401 KAR 31:040, Section 7, Additional Requirement Concerning Nerve and Blistering Agents.

[5] RCRA RD&D Permit Revision 5, February 14, 2014, Tables 3-5 and 3-7.

[6] Ibid., Section 3.2.3.

[7] Kentucky Revised Statutes (KRS) 224.50-130 and 401 KAR 34:350. Treatment of nerve and blister agents.

[8] 40 CFR 260.22.

Recommendation 4-2. BGCAPP should consider working with the CAC/CDCAB Secondary Waste Working Group to pursue a legislative amendment and subsequent modification of Kentucky Department for Environmental Protection regulations such that all treatment-derived wastes that meet the 99.9999 percent standard are delisted (i.e., they will no longer be agent-associated listed wastes); these wastes would then be handled and disposed of according to RCRA requirements—for example, if they demonstrate a RCRA hazardous waste characteristic, they would be managed as RCRA hazardous waste.

RCRA Land Disposal Restrictions

RCRA Land Disposal Restrictions (LDRs) apply to both listed and characteristic wastes and specify either treatment technologies or constituent concentration limits that must be met before a waste can be ultimately disposed of. The federal LDR program does not recognize Kentucky chemical munition or related compounds as listed hazardous wastes, so no specific LDR treatment standards have been established within the federal RCRA program for these wastes. The KDEP has not established state-specific LDR treatment standards for listed chemical GB or VX munition wastes (i.e., N001 or N002).[9] Therefore, since the Kentucky listing for these wastes is only applicable within Kentucky, any hydrolysate or SCWO effluent shipped offsite would only have to respect the LDRs that apply in the state in which the final TSDF is located—for example, that it no longer exhibits a RCRA characteristic (e.g., D002 corrosivity) *and* that it meets any additional treatment requirements for known underlying hazardous constituents prior to final disposal.[10, 11] In addition, under the RCRA LDR regulations, hazardous waste can be stored onsite for more than 1 year only upon a showing that such storage is necessary to facilitate proper future legitimate recycling, treatment, or disposal.[12] This may impact the permissible duration of storage while onsite processes are modified to meet performance criteria or while preparations are being made for offsite shipment.

Any offsite shipment of hydrolysate or SCWO effluent would require characterization of the waste before it could be received by an offsite TSDF. BGCAPP has yet to file its Waste Analysis Plan (Compliance Task 18), which will provide the hazardous waste characteristics and underlying constituents for each waste generated at BGCAPP. The current treatment process planned for BGCAPP requires blending of the agent hydrolysate and the energetics hydrolysate into one tank before entering the SCWO treatment units. However, to simplify waste characterization, BGCAPP indicated that if these hydrolysate wastes need to be sent offsite for treatment, it would normally ship the three types of hydrolysate wastes—GB and VX agent hydrolysates and energetics hydrolysates—separately rather than blending them before shipment.[13] Only hydrolysates already existing in the SCWO blended feed tank would have to be shipped offsite for treatment as a blended hydrolysate.

Finding 4-2. At this time, the RD&D Permit application does not identify the specific characteristics and underlying constituents necessary to determine the applicable Land Disposal Restriction treatment standards for wastes expected to be treated in the SCWO units, including spent decontamination solutions, agent and energetic hydrolysates, or SCWO effluent.

Recommendation 4-3. BGCAPP's Waste Analysis Plan should identify the characteristics and underlying constituents of the spent decontamination solutions, agent and energetic hydrolysates, and SCWO effluent to facilitate identification of appropriate LDR treatment standards to be used by a receiving treatment, storage, and disposal facility.

Structure of RCRA Permit at BGCAPP

By statute and regulation in Kentucky, any applicant for a TSDF permit to treat chemical munition agent wastes and associated compounds must demonstrate that the proposed treatment or destruction technology has been fully proven in an operational facility of scale, configuration, and throughput comparable to the proposed facility (i.e. BGCAPP), or has been demonstrated as effective, within the chemical weapons disposal programs as directed in Congress's establishment of the ACWA program and other applicable federal laws.[14] It must also be demonstrated that such wastes will be destroyed or neutralized at a destruction efficiency of 99.9999 percent under all operating conditions. In addition, the statute and regulations provide that during the occurrence of malfunctions, upsets, or unplanned shutdowns, all quantities of

[9] At this time, Kentucky has generally adopted the federal LDR regulations, but has not received EPA state authorization for all LDR regulations (e.g., the Hazardous and Solid Waste Amendments (HSWA) of 1984 or HSWA phased amendments). Therefore, federal standards would be applied to Kentucky wastes before land disposal for all regulations not yet authorized for the Kentucky program.

[10] It is anticipated in the RCRA RD&D Permit and the BGCAPP *Characterization of Supercritical Water Oxidation (SCWO) Operating Requirements* issued on September 25, 2014, that before the energetic hydrolysate is released to the hydrolysate storage area, it will have already met the RCRA LDR treatment requirements for deactivation so that the waste no longer demonstrates the reactive characteristic.

[11] "Underlying hazardous constituents" do not cause the waste to exhibit a characteristic, but the LDR regulations require underlying hazardous constituents to meet the numeric treatment levels enumerated in the Universal Treatment Standards to be eligible for land disposal (401 KAR 37:040, Section 9/40 CFR 268.48).

[12] 401 KAR 37:050.

[13] Discussions at the Hydrolysate Committee January 2015 meeting in Lexington, Kentucky.

[14] What was originally the Assembled Chemical Weapons Assessment program became the Program Manager for Assembled Chemical Weapons Alternatives and, finally, the Program Executive Office for Assembled Chemical Weapons Alternatives.

any of the chemical munitions wastes shall be contained, reprocessed, or otherwise controlled so as to ensure that the required destruction efficiency is attained.

As indicated previously, BGCAPP plans to operate under a RCRA RD&D permit for the treatment of all GB munitions. An RD&D permit was chosen because the technologies planned for use in the hydrolysis of agent, coupled with the SCWO treatment of the resulting blended agent and energetic hydrolysates, are a first-of-a-kind (FOAK) application of these technologies. As previously discussed, RD&D permits provide more flexibility, still within the regulatory process, to address any technical problems. This structure provides BGCAPP with the flexibility to make operational and infrastructure changes to overcome technical problems and continue onsite hydrolysate processing than would a conventional RCRA TSDF permit. The BGCAPP RD&D permit limits treatment, storage, or disposal to only the types and quantities of hazardous waste that KDEP believes are necessary to determine the efficacy and performance of the FOAK technologies being employed at BGCAPP. It only applies to the destruction of GB munitions. The RD&D permit provides for construction of and operation of the pilot plant for 1 year. It is anticipated that all of the GB munitions will be treated within the first year of the RD&D permit; however, the RD&D permit application provides that should operations take longer than the anticipated 1 year, BGCAPP may seek additional time (up to 3 years total).

BGCAPP plans to file for a Kentucky Hazardous Waste Part B permit to treat the VX munitions stored at BGAD. BGCAPP intends to submit this permit modification application to KDEP at least 2 years before the end of pilot testing under the RD&D permit in order to allow KDEP adequate time to evaluate and eventually approve the permit based on performance tests conducted during the RD&D pilot testing.[15]

The disposal of the GB agent and energetics hydrolysates at an offsite TSDF is not included in the current RD&D permit or in the latest permit modification application.[16] As noted by BGCAPP and KDEP, any modification to eliminate the use of the SCWO treatment unit and instead ship hydrolysates offsite for further treatment would negate the RD&D permit, since BGCAPP would no longer meet the definition of a FOAK pilot treatment facility.[17] If SCWO treatment of the hydrolysate is not possible, and the RD&D permit is negated, BGCAPP would have to request a separate Kentucky Hazardous Waste Part B permit to continue treatment of the GB munitions. However, if only the WRS underperforms or fails to perform, offsite shipment of SCWO effluent may not affect the RD&D nature of the current permit but a modification to the RD&D permit would still be necessary.

According to KDEP, the modification to the BGAD Part B RCRA permit that would be required if SCWO was no longer an option would most likely be a Class III modification.[18] According to KDEP regulations, a determination on a Class III permit modification must be made within 365 days.[19] However, the 365-day duration for completion of a Class III permit modification could be interrupted by several occurrences—namely, those that:

- The time necessary for BGCAPP to respond to KDEP notices of deficiency (i.e. the applicant has 45 days to respond to each notice of deficiency);
- The time for 60-day notice of a public hearing and then 60 days from the date of any public hearing or meeting on the application to allow the Kentucky Environmental and Public Protection Cabinet to consider the public comments (i.e., for as long as 120 days); and
- The 30-day time period allowed for EPA review of and comment on the permit application and another 30 days for EPA review of the draft permit modification (i.e., 60 days).[20]

Therefore, even with no notices of deficiency and no public opposition, a Class III permit modification would take at least 180 days or—more likely—over a year, to be completed.

While either permit modification is being processed (i.e., RD&D or RCRA Part B), KDEP may choose to issue a temporary authorization to, for example, allow BGCAPP to begin construction of facilities that may be needed to facilitate shipments of hydrolysate under the BGAD Part B permit or of SCWO effluent under the RD&D permit. Temporary authorizations typically allow for work such as site preparation, construction, and similar activities to occur while a permit modification is in the approval process. Temporary authorizations to begin construction of ancillary facilities may be useful if it becomes necessary to treat the GB agent and energetics hydrolysates generated at BGCAPP offsite, in order to ensure construction is complete or at least nearing completion at about the same time that the Part B permit modification is granted.

An additional requirement of the BGCAPP RD&D permit is to obtain from Madison County a Host Community Certification letter that infrastructure improvements identified in the Emergency Response Plan are complete and that the Community Liaison position is filled. This letter must be

[15] RCRA RD&D Permit Revision 5, February 14, 2014, Section 2.6.

[16] Hazardous Waste Management Facility Permit (EPA IK DY8-213-820-105, AI 2805) issued on September 20, 2005, and RCRA Research, Development & Demonstration, Permit Revision 5, dated February 14, 2014.

[17] April Webb, KDEP, "BGCAPP RCRA Permitting," presentation to the committee on January 28, 2015.

[18] BGCAPP RCRA Permitting presentation, April Webb, manager, Hazardous Waste Branch, KDEP.

[19] 401 KAR 38:025.

[20] At this time, Kentucky has generally directly adopted the federal permitting regulations but has not received EPA state authorization for all TSDF regulations (e.g., the Hazardous and Solid Waste Amendments of 1984). Therefore, EPA must review all permit applications and modifications.

submitted to KDEP prior to the treatment of any hazardous waste.[21] Any modification to the RD&D permit may require a new Emergency Response Plan and thus a new Host Community Certification.

Any problems with the BGCAPP SCWO or WRS processes would most likely occur during systemization or while operating under the RD&D permit. However, if the severity of any problems makes it appear that offsite transport of the hydrolysate or SCWO effluent may be necessary, BGCAPP would need a Class III Part B permit modification to allow such offsite transport. This process would likely take 1 year or more, as indicated above, including allowing sufficient time for consultations with stakeholders consistent with Recommendations 3-1 and 3-2. Munitions processing could very well be delayed during this period, and would be delayed once hydrolysate storage is full.

As will be discussed in Chapters 6 and 7, BGCAPP has up to 36 weeks of storage for hydrolysate. BGCAPP will therefore have a significant amount of time to either shut down all or some of the SCWO or WRS units pending such investigation, testing, and modifications that are necessary while still processing munitions and storing the hydrolysate. However, it may be that at some point storage capacity will no longer be available and offsite shipment of excess hydrolysate of SCWO effluent may be necessary to facilitate continuous munitions processing pending such modifications.

Finding 4-3. Munitions processing could be delayed for over 1 year due to the regulatory approval process for a Class III RCRA Part B permit modification if problems with slowing or preventing the onsite treatment of hydrolysate cannot be overcome and if it appears that the offsite transport of the hydrolysates will be necessary either as a short-term or permanent solution.

Recommendation 4-4. As a backup plan, BGCAPP should revise its RCRA Part B permit application currently being prepared for the disposal of VX munitions to allow for the possibility of offsite transport of hydrolysates of VX agent, GB agent, and energetics, as well as spent decontamination solution and SCWO effluent, should the SCWO or WRS process be shown to be irreparable.

Recommendation 4-5. As a backup plan, BGCAPP should consult with KDEP concerning whether the RCRA RD&D permit could be modified to allow the temporary offsite transport of GB hydrolysate (i.e., until the SCWO can be brought back on line) or for the temporary or permanent offsite transport of SCWO effluent should the WRS process be shown to be irreparable.

Recommendation 4-6. BGCAPP should consider obtaining a temporary authorization for planning, site preparation, preconstruction, and similar activities for the siting and construction of offsite shipment infrastructure while BGCAPP is still operating under the RD&D permit.

On p. 49 of its 2008 report, the National Research Council (NRC) pointed out as follows:

> On the basis of discussions with state regulators, Mitretek concluded that if offsite shipment of hydrolysate is adopted, neither BGCAPP nor [the Pueblo Chemical Agent Destruction Pilot Plant] would be allowed to begin operations until an appropriate TSDF had been selected and a contract for receipt of the waste was in place. (Bizzigotti et al., 2006)

Even given the necessary permit modification to ship hydrolysates, SDS, or SCWO effluents offsite, no recipient TSDF has yet been identified to accept and treat these wastes. Identifying and contracting with an appropriate recipient TSDF takes time. Any recipient TSDF must have characterization data for the wastes it is to process to demonstrate that it can accept the wastes for treatment under its current RCRA permit. If the BGCAPP wastes are outside of a recipient TSDF's normal waste acceptance criteria, the TSDF may have to process a RCRA permit modification itself. This could take months or longer if there is public opposition. Because federal contracting requirements must be satisfied, establishing a contracting arrangement with a recipient TSDF could also take months as well. Much like obtaining anticipatory permit modifications to allow shipment of hydrolysate, SDS, or SCWO effluent offsite should it prove necessary, it would be desirable for BGCAPP to identify and possibly contract with a TSDF to be ready to accept these wastes, should it prove necessary.

Finding 4-4. The process of identifying and establishing a contract with an appropriate TSDF to receive BGCAPP GB and VX treatment wastes—including hydrolysates, SDS, and SCWO effluent—could take months. This would prevent the expeditious implementation of any decision to ship wastes offsite for disposal and could delay the overall munitions destruction mission.

Recommendation 4-7. As soon as possible, BGCAPP project management should identify at least one offsite TSDF that is approved to accept BGCAPP wastes. It should then establish the necessary mechanisms to quickly contract with the identified TSDF(s).

OTHER ENVIRONMENTAL REQUIREMENTS

Toxic Substances Control Act

The M55 rocket shipping and firing tubes (SFTs) contain polychlorinated biphenyls (PCBs). PCB disposal is regulated under the Toxic Substances Control Act (TSCA). The PCB-

[21] BGCAPP, RD&D Permit, Compliance Schedule, Task 26.

contaminated SFTs will exit BGCAPP along with uncontaminated rocket motors separated from the nonleaking agent-filled warheads for treatment and disposal at an offsite facility permitted under TSCA to receive these wastes. However, as part of the wastes generated during treatment of rockets with leaking warheads, PCB-contaminated SFTs will be processed in the energetics batch hydrolyzer.[22] The BGCAPP RD&D permit modification application indicated that a TSCA permit application has been prepared and submitted to EPA Region 4 for management of PCB-contaminated SFTs as part of the M55 leaker campaign.[23] On the assumption that all PCB contained in the SFTs processed in the energetics batch hydrolyzer are transferred to the hydrolysate, a total PCB concentration per energetics hydrolysis batch of 44 ppm has been calculated. This concentration is below the 50 ppm limit at which TSCA regulation (e.g., permitting) comes into play for onsite SCWO or offsite treatment of this hydrolysate.[24]

Clean Air Act

The gas streams exiting the SCWO unit are expected to be very low in total hydrocarbons (less than 1 ppm). The CO concentration has been demonstrated during the system demonstration programs to be consistently less than 2 ppm, and particulates in the offgases have been less than 4 mg/dry standard cubic meter (DSCM). Cadmium (Cd) and lead (Pb) are less than 0.015 mg/DSCM, and antimony (Sb), arsenic (As), beryllium (Be), and chromium (Cr) are less than 0.045 mg/DSCM. Most of these values are at or below the lower limit of detection of the measurement method, and are below levels commonly found in ambient air, but are not intended to serve as permitting target levels for SCWO gas stream exhaust.[25] The BGAD Title V Air Permit prepared to comply with Clean Air Act requirements was amended to include BGCAPP as a source; the BGCAPP hydrolysate tanks (which are equipped with carbon adsorber systems), the aluminum precipitation and filtration building heating, ventilation, and air conditioning (HVAC) filters system exhaust, the SCWO process building HVAC filter system exhaust, and various SCWO chemical tanks were included as insignificant activities that have to comply with identified applicable regulations. Of those insignificant activities associated with the SCWO, only the SCWO HVAC filter system exhaust must meet a particular applicable regulation: namely, the KDEP requirements for opacity and particulate emissions.[26]

[22] RCRA Research, Development & Demonstration Permit Revision 5, February 14, 2014, Section 3.2.3.5.

[23] Ibid., Section 1.2.3.1.

[24] Battelle Calculation Continuation Sheet, *Concentration of PCBs in one EBH batch*, no date.

[25] RCRA Research, Development & Demonstration Permit Revision 5, February 14, 2014, Section 3.2.3.4.

[26] 401 KAR 59:010, Sections 3(1) and 3(2).

Water Withdrawal Requirements under State Law

Should the WRS function be degraded or cease operations, the recycled reverse osmosis permeate water from the WRS would not be available for SCWO operations. If that were the case, BGCAPP indicated that in order to continue to supply the SCWO processes with quench water it would have to draw replacement water from other available plant sources. However, replacing the reverse osmosis permeate water from the WRS could require increasing BGAD's water withdrawals from Lake Vega. The Final Environmental Impact Statement (FEIS), dated December 2002, found that process water requirements for plant operations for all four technology alternatives, including the neutralization followed by SCWO treatment technology, were within the capacity of Lake Vega.[27] The FEIS found that agent neutralization followed by SCWO treatment technology would require approximately 6 million gallons of process water annually, and that no present or planned activities were identified that would result in withdrawals in excess of the quantity specified in the water permit issued to BGAD by KDEP: a monthly average of 500,000 gal. If necessary, the BGAD water withdrawal permit could be modified to accommodate the increased demand for water, and the 500,000 gal storage tank could provide a short-term supply of process water. However, in the event of an extreme and prolonged drought, the FEIS assumed that agent neutralization operations would be halted before reduced water supply availability jeopardized plant safety.

The review of an application for a revision of a water withdrawal permit will take 90 calendar days after receipt of an administratively complete permit application, excluding any time necessary for (1) response to notices of deficiency; (2) litigation; (3) public hearing or public comment period on a draft or proposed permit; or (4) any federal, state, or local agency comment period or Kentucky Energy and Environment Cabinet requests for additional information. The issue date of the permit may be as much as 3 years in advance of the effective date, so any modification granted can be held for 3 years and then allowed to lapse once the water is no longer needed as backup for the process water from the WRS. However, should the existing water withdrawal limits, along with the current water storage, not be able to supply sufficient water for SCWO requirements, munition processing may have to be slowed or stopped until the permit modification is granted so as to not jeopardize plant safety.

Finding 4-5. If it appears that obstacles to WRS issues cannot be overcome, or overcome before the existing process water storage capacity is expended and additional water withdrawals that exceed the current permitted volumes become necessary, munitions processing would likely be delayed for

[27] Final Environmental Impact Statement, 2002, Destruction of Chemical Munitions at Blue Grass Army Depot, Kentucky, December.

over 90 days based on the regulatory approval process for a Kentucky water withdrawal permit modification.[28]

Recommendation 4-8. BGCAPP should determine the potential shortfall of process water if reverse osmosis permeate water is not able to be reused and ensure there is sufficient plant water from BGAD. If there are insufficient plant water supplies available to maintain BGCAPP's proposed throughput schedule, BGCAPP should, as a backup plan, modify its Kentucky water withdrawal permit to allow for sufficient water sources to maintain operations of the plant without the reverse osmosis permeate water.

NATIONAL ENVIRONMENTAL POLICY ACT REQUIREMENTS

As indicated in the NRC report *Review of Secondary Waste Disposal Planning for the Blue Grass and Pueblo Chemical Agent Destruction Pilot Plants* (2008), BGCAPP prepared and issued an environmental impact statement (EIS), under the National Environmental Policy Act (NEPA), that covers the construction and operation of the facility. The draft EIS for the destruction of chemical agents and munitions stored at BGAD was released for public comment in May 2002. It considered the environmental impacts of no action, incineration, two neutralization technologies, and electrochemical oxidation. The FEIS incorporated all comments. The FEIS did not address offsite shipment of hydrolysate or SCWO effluent. The Record of Decision, issued February 27, 2003, does not consider offsite shipment of hydrolysate or SCWO effluent and, in fact, finds that the quantity of hazardous liquid wastes is expected to be small to nonexistent (due to recycling) for all four treatment alternatives considered. According to NEPA, if an existing EIS or environmental assessment does not adequately cover the new proposed action, another EIS or environmental assessment would have to be prepared that would either result in a finding of "no significant impact" or a requirement to prepare further NEPA documentation (40 CFR Section 1502.9(c)(1)). Preparing this documentation could delay munitions disposal if hydrolysate storage fills before hydrolysate can be shipped, resulting in a slowdown or halt to plant operations.

The importance of addressing the potential for offsite shipment in the BGCAPP environmental documentation is shown by the court case *The Sierra Club et al. v. Dr. Robert M. Gates, Secretary of Defense, et al.*, brought in the U.S. District Court of the Southern District of Indiana, Terre Haute Division. The plaintiffs in this case wanted to stop the government from continuing the shipment of VX hydrolysate from the Newport Chemical Agent Disposal Facility (NECDF) to Veolia's incineration facility in Port Arthur, Texas. The court ruled that the FEIS and Record of Decision for its Chemical Stockpile Disposal Program indicated that a site-specific NEPA review, including the preparation of either an EIS or an environmental assessment, would be conducted for each chemical stockpile storage location.

The court found that the 1998 FEIS for the pilot test of its VX neutralization plan at NECDF evaluated two alternatives: (1) take no action, and (2) use the proposed process. In response to NRC studies and the September 11, 2001, attacks, the Army published its Final Environmental Assessment in July 2002 considering the VX destruction process at NECDF (ORNL, 2002). In that assessment, the Army compared two alternatives: (1) taking no action and (2) disposing of the hydrolysate at an offsite TSDF. The assessment made no findings concerning a specific TSDF. Rather, it suggested that the appropriate analysis would be performed after a TSDF had been chosen to dispose of the hydrolysate. The Army issued a Final Finding of No Significant Impact on October 28, 2002 (CMA, 2002).

After performing a detailed analysis and engaging in discussions with the Veolia facility, EPA approved the offsite treatment option in 2006. The Army issued a record of environmental consideration (REC) in April 2007 for the proposed shipment of caustic wastewater (i.e., VX hydrolysate) to the Veolia facility and found that the proposed action qualified for a categorical exclusion the Army has for the routine management of hazardous materials or operations involving hazardous waste.[29] The Army used the existing NEPA documentation, including previous FEIS and Final Environmental Assessment documents, to support the REC it had issued in April 2007. The Army then issued another REC in June 2007 in response to a letter from the plaintiffs about the safety and environmental impacts of the proposed shipment of VX hydrolysate. The Army used the same rationale for not performing an EIS or environmental assessment as was used in the earlier REC—that is, that the VX hydrolysate was classified as hazardous waste and came under the Army's categorical exclusion for routine treatment and handling of hazardous waste originating at its facilities.[30]

In the end, the court ruled that it may not substitute its own judgment for that of an agency. Rather, it must defer to that agency's factual findings when deciding whether the environmental impacts of its actions are significant. The court agreed with the Army's contention that it did not need to supplement either its 1998 FEIS or its 2002 Final Environmental Assessment. It also ruled that the Army did

[28] It is estimated that BGCAPP can continue to operate the SCWO for a maximum of 1.5 days without the WRS. NRC Request Update, NRC#08, received February 16, 2015.

[29] A categorical exclusion is defined as actions that normally do not require an environmental assessment or EIS, and the Army has determined that they do not individually or cumulatively have a substantial effect on the human environment. From Appendix B of 32 CFR Part 651 (AR 100-2), Environmental Analysis of Army Actions.

[30] 32 CFR 651, Appendix B, § (h)(4). See also Memorandum for U.S. Army Chemical Materials Agency Commanders, et al., June 25, 2007, U.S. Army Chemical Materials Agency, Guidance for Development of Site-Specific Plans for Shipment of Chemical Agent Contaminated Secondary Waste.

not need to provide an additional comment period when the alternative option was proposed. The court found that the administrative record reflected that the original NEPA documents considered the cases of onsite hydrolysate treatment and the shipment of hydrolysate to a permitted TSDF offsite. The court also found that, because the government had already taken the necessary "hard look" at the nature of the hydrolysate and correctly determined it to be hazardous waste, a secondary decision to switch to another permitted TSDF did not raise the requirement for a supplemental EIS or environmental assessment.[31,32]

Any permit modification to ship hydrolysates offsite from BGCAPP would require a determination whether BCAPP's current NEPA documentation is adequate or if supplemental documentation is needed. Unlike the NECDF experience, the BGCAPP NEPA documentation did not include the alternative action for shipping hydrolysates offsite. Therefore, BGCAPP may find it necessary to take a "hard look" at the nature of the hydrolysates and other wastes it will produce and potential impacts of shipping them offsite for disposal. It would then have to determine whether offsite shipment would be covered by the existing NEPA documentation, whether such shipments would come under the categorical exclusion, or if it would be necessary to generate supplemental NEPA documentation in support of offsite shipment.

Finding 4-6. If the supercritical water oxidation and water recovery systems underperform, or fail to perform, to the extent that consideration of offsite disposal becomes necessary, it may also be necessary to conduct environmental analyses under the National Environmental Policy Act. This has the potential to significantly delay offsite shipment.

Recommendation 4-9. BGCAPP project management should begin immediately to determine whether its current NEPA documentation is adequate to support offsite hydrolysate or other waste shipment or if new documentation is needed. This documentation could possibly include a new Final Environmental Assessment, a new Final Environmental Impact Statement, or a document to support categorical exclusion if one applies. This determination should be completed while BGCAPP is still operating under its RCRA RD&D permit. BGCAPP management should also determine how much time would be required to prepare any new NEPA documentation required and pre-file it.

[31] Courts consistently have held that, at a minimum, NEPA imposes a duty on federal agencies to take a "hard look at environmental consequences" (*Natural Resources Defense Council v. Morton*, 458 F.2d 827, 838 (D.C. Cir., 1972)).

[32] The Army switched from shipping to the DuPont treatment facility in Deepwater, New Jersey, to shipping to the Veolia incinerator facility in Port Arthur, Texas.

ORGANISATION FOR THE PROHIBITION OF CHEMICAL WEAPONS

As indicated in Chapter 1, the United States is a signatory to the Chemical Weapons Convention (CWC), which is overseen by the Organisation for the Prohibition of Chemical Weapons. Under the CWC, BGCAPP is subject to onsite monitoring by representatives of the Organisation for the Prohibition of Chemical Weapons. In the United States, the treaty is administered through the U.S. Department of State. Destruction, under the CWC, is the process by which chemicals are converted in an essentially irreversible way to a form unsuitable for production of chemical weapons and which irreversibly renders munitions and other devices unusable as such.

In addition to GB and VX, the CWC also monitors compounds listed on CWC Schedules 1 and 2. These compounds include dual-use chemicals that have legitimate industrial uses but can also be used to manufacture chemical agents or their precursors. Schedules 1 and 2 chemicals may be present in hydrolysate as breakdown products of the chemical agents, as discussed in Chapter 2.

VX hydrolysate may contain EA 2192, which is a Schedule 1A chemical. Thus, EA2192 may be present in the VX hydrolysate fed to SCWO. BGCAPP plans to take samples from neutralization batches of VX agent hydrolysate, and VX-associated energetics hydrolysate will be collected and analyzed by the onsite laboratory to ensure batches meet the clearance requirement for EA2192, currently anticipated to be 2.3 ppm, prior to release of the batch from the Munitions Demilitarization Building.

In addition, some of the phosphorus-containing byproducts in hydrolysates are considered to be Schedule 2 compounds and must be irreversibly destroyed to meet the requirement of the CWC. In VX hydrolysate, these compounds include ethyl methyl phosphonic acid (EMPA), methyl phosphonic acid (MPA), and O-ethyl methyl phosphonothioic acid (EMPSH). In GB hydrolysate, isopropyl methyl phosphonic acid (IMPA) and MPA are Schedule 2 compounds. Note that organic acids are present in the caustic hydrolysate solutions as their conjugate bases. Other organophosphorus compounds may be present as well due to impurities introduced during agent production and miscellaneous degradation reactions during storage.[33]

Finding 4-7. The Chemical Weapons Convention (CWC) requires that all destruction facilities are subject to certain declaration, reporting, and inspection requirements and verification of full destruction. Destruction means a process by which chemicals are converted in an essentially irreversible way to a form unsuitable for production of chemical weapons, and the CWC requires that the destruction process must be verifiable.

[33] Ibid.

Finding 4-8. Shipment of hydrolysate containing EA2192, a Chemical Weapons Convention (CWC) Schedule 1A chemical, and Schedule 2 chemicals may require verification under the CWC of full destruction.

Recommendation 4-10. During systemization testing with hydrolysate surrogates, BGCAPP project management should work with the Kentucky Department for Environmental Protection and the Organisation for the Prohibition of Chemical Weapons as necessary to demonstrate that the BGCAPP is effective in treating hydrolysates such that any shipments to offsite disposal facilities will have met all CWC requirements.

REFERENCES

DEP (Department for Environmental Protection). 2005. *Standards for Owners and Operators of Hazardous Waste Storage, Treatment and Disposal Facilities*. Title 401, Chapter 34 of the Kentucky Administrative Regulations (KAR). http://www.lrc.ky.gov/kar/title401.htm.

Bizzigotti, G.O., M.A. Berger, T.C. Cain, D.J. Cleaves, C. Gomolka, E.W. Hughitt, D.M. Ligon, P.K. McDonald, R.P. Rhoads, and A.R. Wusterbarth. 2006. *Analysis of Off-Site Treatment of Hydrolysates from Chemical Agent Destruction Pilot Plants*. Mitretek Report MTR 2006-22, October. Falls Church, Va.: Mitretek Systems.

NRC (National Research Council). 2008. *Review of Secondary Waste Disposal Planning for the Blue Grass and Pueblo Chemical Agent Destruction Pilot Plants*. Washington, D.C.: The National Academies Press.

ORNL (Oak Ridge National Laboratory). 2002. *Final Environmental Assessment Accelerated Neutralization of Chemical Agent and Off-Site Shipment of Liquid Process Effluents at the Newport Chemical Agent Disposal Facility*. Oak Ridge, Tenn.

CMA (U.S. Army Chemical Materials Activity). 2002. *Draft Revised Final Finding of No Significant Impact* (FONSI), Accelerated Neutralization of Chemical Agent and Off-Site Shipment of Liquid Process Effluents at the Newport Chemical Agent Disposal Facility. www.pmcd.army.mil/fndocumentviewer.aspx?DocID=003675301.

5

Transportation of Chemical Materials

This chapter addresses the risks of transporting hazardous materials, including hydrolysate. It first describes the federal regulations that define and govern the reporting of heavy truck accidents (i.e., crashes) and hazardous material incidents. Next it reviews historical data for the offsite transportation of hydrolysate and similar materials from chemical demilitarization sites. Finally the chapter addresses the risks of transporting hydrolysate. These risks include

- Those associated with heavy truck crashes, regardless of cargo;
- Those associated with transportation of hazardous material cargo in general; and
- Those unique to transportation of hydrolysate.

DEPARTMENT OF TRANSPORTATION REGULATIONS

The U.S. Department of Transportation (DoT) has the primary responsibility for protecting and enhancing the safety, adequacy, and efficiency of the national transportation system and related services. It comprises 11 individual operating administrations. Of these, the Pipeline and Hazardous Materials Safety Administration is responsible for the regulations governing hazardous materials, including their classification into one of nine classes and the associated vehicle placarding, packaging, and other requirements (49 CFR 171-180).

To date, all hydrolysate shipments have been by truck. Accordingly, this chapter emphasizes truck transport on public roads and highways. There may also be an option to transport hydrolysate by rail. Rail transport would present risks to the public similar to truck transportation: for example, direct physical impact at a crash scene and the release of the cargo to the environment, possibly exposing the public. This section introduces the concepts of "reportable crash," "reportable incident," and the first step of hazard classification, which dictates the subsequent regulations that must be adhered to.

The Federal Motor Carrier Safety Administration maintains a database on serious truck and bus crashes. A "DoT-reportable" crash is reported to the Federal Motor Carrier Safety Administration if it has the following elements (49 CFR 390.5):

> A truck having a gross vehicle rating of over 10,000 pounds or any vehicle that displays a hazardous material placard,
>
> and
>
> A crash occurs while the vehicle involved is operating on a roadway that is normally open to the public, and results in
>
> —Fatality; or
> —An injury requiring medical treatment away from the crash scene; or
> —The towing of any motor vehicle disabled in the crash.

A reportable hazardous material incident is defined and reported to the National Response Center, if as a direct result of a hazardous material (49 CFR 171.15),

- A person is killed,
- A person receives an injury requiring admittance to a hospital,
- The general public is evacuated for 1 hour or more, or
- A major transportation artery is shut down for 1 hour or more.

The incident is reported separately to the Pipeline and Hazardous Materials Safety Administration if, in addition to the circumstances in 49 CFR 171.15, the incident results in (49 CFR 171.16) the following:

- An unintentional release of a hazardous material or hazardous waste, or
- A specification cargo tank of 1,000 gallons or more containing hazardous material suffers damage to the lading retention system or to a system intended to protect the lading retention system.

TABLE 5-1 Historical Data on the Shipment of Hydrolysates and Neutralent

Parameter	Operation Swift Solution GB Hydrolysate (JPEOCBD, 2014)	HD Hydrolysate (JPEOCBD, 2014)	VX Hydrolysate (JPEOCBD, 2014)	GB Bomblet Destruction (RCMD, 2014)
Origin	BGAD	APG	NECD	RMA
Destination	Veolia TSDF, Port Arthur, Tex.	DuPont TSDF, Deepwater, N.J.	Veolia TSDF, Port Arthur, Tex.	Safety-Kleen TSDF, Deer Park, Tex., or APG, Md.
Number of shipments	2	Approximately 1,450	424	2/1
One-way mileage	1,140	49	1,011	1,032/1,705
Total shipment mileage	2,280	Approximately 69,580	428,664	3,769
DoT label and marking (flash point <200ºF if applicable)	Class 8, Packing Group II, waste corrosive liquid, basic organic, n.o.s., UN3267, RQ (sodium hydroxide)	Class 8, Packing Group II, corrosive liquids, n.o.s. (thiodiglycol + 5% NAOH solution + D16), UN1760	Class 8, Packing Group II, waste corrosive liquid, basic organic, n.o.s., UN3267, RQ (sodium hydroxide)	Uncertain
DoT reportable accidents (crashes)	None reported	None reported	None reported	None reported
Incidents	None reported	None reported	None reported	None reported
Nonreportable crashes (fender benders)	None reported	None reported	None reported	None reported

NOTES: n.o.s., not otherwise specified; JPEOCBD, Joint Program Executive Office for Chemical and Biological Defense; VX, a nerve agent; HD, distilled mustard agent; TSDF, treatment, storage, and disposal facility; RCMD, Recovered Chemical Material Directorate.

It should be noted that terminology differs across different DoT documents. The regulation defining "accident" (49 CFR 390.5) uses "occurrence" instead of "crash." Further, "DoT-reportable" usually includes "accident." However, many DoT documents use "crash" rather than "accident" to clearly indicate the presence and involvement of physical forces (e.g., DoT, 2014). It should also be noted that an incident may involve a hazardous material release without involving a crash. In this chapter the committee uses "accident" in the phrase "DoT-reportable accident" and "crash" elsewhere unless it is quoting a document.

To date, hydrolysate and similar liquids have been designated as Class 8 corrosive materials. A Class 8 hazardous material is defined as a liquid or solid that causes either (1) the destruction of the full thickness of human skin within a specified time period or (2) a specified corrosion rate for steel or aluminum (49 CFR 173.136).[1] The destruction rate defines the packing group for the material, the groups being I, II, or III. The hazardous material class and the associated packing group dictate a number of important DoT requirements, such as the selection of equipment and the procedures used when conducting inspections. While DoT regulations do not explicitly define "corrosive" using a pH value, the U.S. Environmental Protection Agency (EPA) defines hazardous waste as corrosive if either (1) it is a liquid with pH <2 or >12.5 or (2) a liquid that corrodes steel at a rate of >0.250 in./yr at a test temperature of 130°F (55°C). The EPA and DoT definitions of corrosive are frequently confused.

THE HISTORICAL TRANSPORTATION OF CHEMICAL MUNITION MATERIALS

Historical data are shown in Table 5-1 for the offsite transportation of GB, HD, and VX hydrolysate from, respectively, the Blue Grass Army Depot (BGAD) during Operation Swift Solution, Aberdeen Proving Ground (APG), and the Newport Chemical Depot (NECD), as well as the Explosive Destruction System (EDS) neutralent from destruction of GB bomblets at Rocky Mountain Arsenal (RMA) in 2001. All of these shipments were by truck. The composition of most materials in those shipments is given in Table 5-2; they can be compared to the anticipated composition of hydrolysates from the Blue Grass Chemical Agent Destruction Pilot Plant (BGCAPP) shown in Tables 2-1 through 2-6. The data in Tables 5-1 and 5-2 show that over 500,000 miles have been accumulated shipping materials similar to the hydrolysate that is anticipated to be generated at BGCAPP without a leak or even a minor crash.

[1] Examples of Class 8 materials are hydrochloric acid, nitric acid, sulfuric acid at a concentration of >51 percent, and solid sodium hydroxide, all commonly transported materials.

TABLE 5-2 Comparison of Historical Hydrolysate and Neutralent Contents

Parameter	APG HD Hydrolysate[a]	NECD VX Hydrolysate[b]	RMA GB Bomblet Neutralent[c]
Primary active ingredient	Hot water and NaOH	Water and NaOH	Monoethanolamine
Water (wt%)	88-97	71-91.7	51.7-56.2
Approximate pH	12.4	12.5-14	12
Thiodiglycol (TDG) (ppm)	52,250		NA
Isopropyl methylphosphonate (IMPA) (ppm)			3,400-5,000
Diisopropyl methylphosphonate (DIMP) (μg/L)			18,000-27,400
Sodium 2-(diisopropylamino) ethylthiolate (%)		<11	
Sodium ethylmethyl phosphonate (%)		<10	
Sodium methyl phosphonate (%)		<2	
Diisopropylamine (%)		<4	
1,4-Dithiane (ppm)	1,371		
1,4-Oxathiane (ppm)	734		
1,2-Dichloroethane (ppm)	181		
Total organic carbon	27,875 mg/L	<12 %	
Total suspended solids	8,676 mg/L	<1.0 %	
Benzene	319 ppb		1,300-2,850 μg/L
Chloroform	329 ppb		ND-21.6 μg/L
Dichloromethane (μg/L)			ND-97.1
Toluene	58 ppb		369-810 μg/L
Vinyl chloride (ppm)	12		
Ammonia (ppm)		<500	
Arsenic	2,297 ppb	<5 ppm	<200 μg/L
Barium (ppm)		<100	
Cadmium	95 ppb	<1 ppm	6.81-10 μg/L
Chromium	1,639 ppb	<5 ppm	445-770 μg/L
Copper	6,515 ppb	<1 ppm	9,030-18,200 μg/L
Lead	1,377 ppb	<5 ppm	63-237 μg/L
Mercury	164 ppb	<0.2 ppm	0.1-1 μg/L
Iron (ppm)	2,161	<5	
Selenium (ppm)		<1	
Silver (ppm)		<5	
Zinc	3,811 ppb	<10 ppm	23,100-38,300 μg/L
Explosives in liquids (μg/L)			<1,000

[a] Aberdeen Chemical Agent Disposal Facility shipment analysis data for shipments between June 14, 2004, and February 9, 2005, provided by Bill Steedman.
[b] December 12, 2006, Waste Characterization Sheet.
[c] Laurence Gottschalk, director, Recovered Chemical Materiel Directorate, "Recovered Chemical Materiel Directorate (RCMD) Bomblet Destruction Campaign at Rocky Mountain Arsenal," presentation to the committee on July 30, 2014.
NOTE: NaOH, sodium hydroxide (caustic); ND, not detected; ppm, parts per million; ppb, parts per billion.
SOURCES: Information provided to the committee on July 30, 2014.

Finding 5-1. The accumulated mileage for the historical shipment of hydrolysate and similar materials is dominated by the shipment of hydrolysate from NECD to the Veolia treatment, storage, and disposal facility in Port Arthur, Texas. The shipments from the Newport Chemical Depot, the Blue Grass Army Depot, Aberdeen Proving Ground, and Rocky Mountain Arsenal were free from even minor crashes and from any leaks of hydrolysate or similar fluids.

IDENTIFICATION OF THE RISKS OF TRANSPORTING HYDROLYSATE

Risk is the combination of the likelihood of a specified hazard being realized and the consequence of that hazard occurring. Likelihood in transportation risk analyses is usually expressed as crashes per mile or crashes per trip. Some of the factors that affect the likelihood are the number of shipments, the distance traveled per shipment, the route characteristics, the carrier, and the transportation mode. Conditional probabilities based on the factors in the consequence calculation are generally included in the likelihood calculation, e.g., the probability of a fire, given a crash, and the probability of fire causing a hazardous material container to fail, given that a fire occurs.[2] The potential consequences include

[2] The hydrolysates will not themselves be flammable, so a simple spill would not result in a fire. Given a large truck, the probability of a fire is only a few percent.

injuries and fatalities not only due to the impact of a heavy truck with a person but also the exposure of members of the public to a released hazardous material. Some of the factors that affect exposure are the dispersion of the material not only due to the material properties such as vapor pressure, but also meteorological dispersion characteristics and the potential presence of a fire. Given an exposure, the health effects vary with the level of toxicity, corrosiveness, and so forth.

Methodologies used to analyze risk differ in their scope. One might entail selecting the most appropriate transportation mode or hazardous materials container and determining the necessary level of detail to meet the purpose of the analysis in question. Another methodology might look at whether the simple release of a hazardous material is the appropriate consequence, or whether dispersion and possible public exposure also need to be considered, factoring in the appropriate conditional probabilities, such as those mentioned above. All risk analyses aggregate to some degree the various factors that produce and affect risk. The extent of aggregation depends on the availability of data and the purpose of the analysis. The resulting analyses may be quantitative, qualitative, or quantitative for some portions and qualitative for others. This chapter does not specify the level of detail that would be appropriate for a BGCAPP quantitative transportation risk analysis (QTRA), or the numerous factors that should be included in the analysis. That is for BGCAPP to determine during the conduct of a QTRA. However, some general requirements for such an analysis are identified at the end of this chapter. Instead, this chapter examines how transportation risk changes with cargo type. Some quantification is provided to help evaluate risks associated with various cargo types.

The risks identified in this section are those associated with the following consequences:

- Fatalities and/or injuries resulting from the impact of a heavy truck with a person, independent of the cargo;
- Fatalities, injuries, and/or economic consequences resulting from the release of hazardous materials; and
- Those unique to hydrolysate.

As previously stated, transporting hydrolysate by rail is an option, but the focus of this chapter is on transportation by truck across public roads and highways. The process for identifying risk for rail transportation would be similar to the process for truck transportation.

Risks Due to a Heavy Truck Crash, Independent of Cargo

The likelihood of a large truck being involved in a serious crash is about 1.2×10^{-6}/mi, or somewhat more than 1 for each million miles traveled. In the event of a serious heavy truck crash, the probability of a fatality is about 1 percent, and that of an injury is about 23 percent, each independent of the cargo being transported (DoT, 2014).

Finding 5-2. There is a low risk of injuries and fatalities resulting from a heavy truck crash, independent of the cargo being transported.

Additional Risks from Carrying Hazardous Materials

On top of the cargo-independent risks resulting from a large truck crash, the risks associated with a release of hazardous materials include injuries, fatalities, and cleanup costs. A recent Transportation Research Board report notes that "hazmat-specific accident rates are usually not available and truck accident rates are often used as a proxy" (TRB, 2013). The reason for this is, in part, that private and public stakeholders (e.g., the Transportation Security Administration) protect data for a variety of reasons, such as maintaining competitiveness and operational security.

The current crash rate for Tri-State Motor Transit Company, a company frequently contracted to transport hazardous materials and the carrier contracted to transport the hydrolysate from NECD, is 0.38 in a million miles (0.38×10^{-6}/mi), or about one-third the rate for heavy trucks in general (DoT, 2015a). This rate applies to the company's entire fleet, not just the hazardous materials portion. The crash rate for the hazardous materials portion of their fleet can be expected to be less than 0.38×10^{-6}/mi owing to the extra qualifications required of drivers who transport hazardous materials and other requirements for hazardous material shipments.

Table 5-3 summarizes highway incidents in 2014 by transport phase. While the number of incidents in transit is only slightly more than about a third of those occurring during loading and unloading, they cause the majority of fatalities and monetary damages. There are over 800,000 highway

TABLE 5-3 Highway Hazardous Material Incident Summary by Transportation Phase in 2014

Transportation Phase	Incidents	Injuries (Hospitalized)	Injuries (Not Hospitalized)	Fatalities	Damages ($)
In transit	3,198	1	30	6	44,036,568
In transit storage	356	1	1	0	1,844,162
Loading	3,245	3	23	0	1,002,237
Unloading	8,391	8	51	1	6,925,600
Total	15,190	13	105	7	53,808,567

SOURCE: DoT, 2015b.

hazardous material shipments each day (DoT, 2004), about 300 million per year. The 15,190 incidents in 2014 represent a very small proportion of hazardous material shipments on highways. Since the contribution of hazardous material risk is small compared to general truck transportation risk, fatalities and injuries related to hazardous material cargos are frequently neglected in the face of the much greater risk of cargo-independent fatalities and injuries.

Finding 5-3. The historical risk of a hazardous material incident during transportation-has been small.

Additional Risks Posed by Transporting Hydrolysate

If hydrolysate is not processed through the supercritical water oxidation (SCWO) system and is shipped offsite, BGCAPP estimates that 221 truckloads of GB hydrolysate, 40 truckloads of VX hydrolysate, 386 truckloads of energetics hydrolysate, and 842 truckloads of condensate from the offgas treatment for the metal parts treater (which would have otherwise been processed through the SCWO) would be required. However, since reverse osmosis (RO) reject water from the water recovery system would no longer be produced, offsite shipments of RO reject would not be required, thus eliminating about 5,700 shipments from those currently planned. Overall, if hydrolysate is shipped offsite for treatment, there would be about 4,200 fewer shipments of waste material from BCAPP in the case of offsite treatment of hydrolysate.[3] The committee believes that additional reductions in offsite shipments are likely in the case of offsite hydrolysate shipment, owing, for example, to the lack of aluminum filter cake to be transported, but those data were not provided. Other shipments required for BGCAPP operations, for example, caustic for the production of hydrolysate, would be essentially unchanged. To put these numbers in perspective, Noblis (2008) estimated the total number of shipments as 9,088 for the case of onsite treatment of hydrolysate.

As discussed in Chapters 4, 6, and 7, if SCWO performs adequately but the water recovery system fails, there is a possibility of shipping SCWO effluent offsite. In this case, the currently planned offsite shipments would change, primarily because the 5,700 RO reject shipments (30 percent of SCWO effluent) would be replaced by the shipment of the entire SCWO effluent, an increase of 70 percent of the SCWO effluent, or about 13,300 shipments. Because it is dilute brine, SCWO effluent would not pose a significant chemical toxicity hazard should a release occur in transport.

The likelihood of an injury or a fatality due to a heavy truck crash, provided above, applies to both the shipments planned for onsite hydrolysate treatment and the shipments that would be necessary with offsite hydrolysate (or SCWO effluent) treatment, as does the very low risk of a hazardous material incident.

Finding 5-4. Offsite hydrolysate transport would decrease the total number of shipments from BGCAPP by a net of about 4,200 shipments, about half the shipments with onsite treatment.

Finding 5-5. Offsite shipment of the entire SCWO effluent would increase the total number of shipments from BGCAPP by about 13,300 shipments, about double the shipments for onsite treatment.

Hydrolysate shipments from NECD and BGAD were subjected to enhanced safety measures as compared with typical hazardous material shipments, including safety inspections every 2 hours or so while on the road (Veolia, 2009). The same safety measures can be expected if hydrolysate is shipped from BGCAPP. Thus, a crash rate less than the 0.38×10^{-6}/mi cited above for a representative hazardous material transportation company can be expected for BGCAPP hydrolysate shipments. Further, a comparison of material compositions of historical hydrolysate shipments shown in Table 5-2 with anticipated compositions of the BGCAPP hydrolysates in Tables 2-1 through 2-6 provides no reason to expect any appreciable change in transportation risk due to the nature of the BGCAPP hydrolysates.

RELATED PRIOR FINDINGS AND RECOMMENDATIONS FROM NATIONAL RESEARCH COUNCIL REPORTS

The 2008 National Research Council (NRC) report *Review of Secondary Waste Disposal Planning for the Blue Grass and Pueblo Chemical Agent Destruction Pilot Plants* (NRC, 2008) contained findings and recommendations relevant to the consideration of offsite transport of hydrolysate from BGCAPP. The 2008 report expressed a concern that hydrolysate-specific risks should be compared quantitatively with the cargo-independent risks from heavy trucks, rather than through a separate, independent qualitative methodology. The report pointed out that it is important to provide quantitative data to address concerns expressed about the prospect of offsite transportation. The 2008 report contained the following finding and recommendations that are applicable to BGCAPP:

> **Finding 6-4.** Some members of the public and state regulators are concerned about the health risks of hydrolysate transport and believe there is a need for emergency planning along the route.
>
> **Recommendation 6-3.** The [Program Manager for Assembled Chemical Weapons Alternatives] should perform a quantitative transportation risk assessment for hydrolysate, including a quantitative assessment of the human health consequences of hydrolysate spills with and without a fire. This

[3] E-mail correspondence between John Barton, chief scientist, BPBG, and Jeff Krejsa, PE deputy site project manager, Compliance, ACWA BGCAPP, on March 11, 2015.

assessment needs to be completed to facilitate discussions with the public and regulators about the hydrolysate offsite shipment alternative.

Recommendation 6-4. The [Program Manager for Assembled Chemical Weapons Alternatives] should prepare a prototypical emergency response plan for hydrolysate shipment, including the possibility of a fire or the occurrence of natural disasters such as floods. This plan would be the starting point for setting contractual requirements for the [treatment, storage, and disposal facility] and the shipper. The prototype needs to be completed to facilitate discussions with the public and regulators about the hydrolysate offsite shipment alternative.

Finding 6-8. The experience to date with the offsite shipment and treatment of mustard and nerve agent hydrolysates from the Aberdeen and Newport Chemical Agent Disposal Facilities indicates that offsite transport and disposal of these materials is a safe and technically viable course of action.

Finding 5-6. The findings and recommendations cited above from *Review of Secondary Waste Disposal Planning for the Blue Grass and Pueblo Chemical Agent Destruction Pilot Plants* (NRC, 2008) are still relevant.

Recommendation 5-1. The transportation-related recommendations in *Review of Secondary Waste Disposal Planning for the Blue Grass and Pueblo Chemical Agent Destruction Pilot Plants* (NRC, 2008) should be followed.

There is enough experience from transporting hydrolysate from BGAD, APG, and NECD to be able to perform a reasonable QTRA in the near term. The main piece of information still lacking for any analysis is the receiving treatment, storage, and disposal facility (TSDF). Procurement regulations may prevent the timely selection of a receiving TSDF. One option to address this is to perform a QTRA for one or more representative TSDFs. While the committee recognizes that this would entail a significant amount of work, it believes this approach would have several benefits, including these:

- Illustrating the impact of route selection on risk,
- Quantifying the relative contributions to risk from both the cargo-independent and the hydrolysate release scenarios,
- Quantifying the overall magnitude of the risk associated with hydrolysate transportation sufficiently for regulators and stakeholders to evaluate it, and
- Providing input for emergency response planning processes.

REFERENCES

DoT (Department of Transportation). 2004. Crashes Involving Trucks Carrying Hazardous Materials, Federal Motor Carrier Safety Administration, FMCSA-RI-04-024, May 2004. Available at http://www.fmcsa.dot.gov/facts-research/facts-figures/analysis-statistics/fmcsa-ri-04-024.htm.

DoT. 2014. *Large Truck and Bus Crash Facts–2012*, Federal Motor Carrier Safety Administration, FMCSA-RRA-14-004, June 2014. Available at http://www.fmcsa.dot.gov/safety/data-and-statistics/large-truck-and-bus-crash-facts-2012.

DoT. 2015a. Company Snapshot: Bed Rock Inc. Accessed February 6, 2015. Available at safer.fmcsa.dot.gov.

DoT. 2015b. PHMSA Incident Statistics. Accessed March 11, 2015. Available at www.phmsa.dot.gov/hazmat/library/data-stats/incidents.

JPEOCBD (Joint Program Executive Office for Chemical and Biological Defense). 2014. *Offsite Hydrolysate Transport and Disposal Case Studies*. Provided to the committee on July 30, 2014.

Noblis. 2008. *Offsite Disposal of ACWA Hydrolysates*. NTR 2008-61129. Falls Church, Virginia.

NRC (National Research Council). 2008. *Review of Secondary Waste Disposal Planning for the Blue Grass and Pueblo Chemical Agent Destruction Pilot Plants*. Washington, D.C.: The National Academies Press.

RCMD (Recovered Chemical Material Directorate). 2014. *Bomblet Destruction Campaign at Rocky Mountain Arsenal.* Provided to the committee by Larry Gottschalk on July 30, 2014.

TRB (Transportation Research Board). 2013. *Hazardous Materials Transportation Risk Assessment: State of the Practice,* Visual Risk Technologies, Inc., Hazardous Materials Cooperative Research Program Report 12. Washington, D.C.

Veolia Environmental Services. 2009. *Transportation Safety and Emergency Response Plan for the Shipment of Operation Swift Solution Caustic Wastewater*.

6

Hydrolysate Treatment Criteria for Success and Decision Framework

As described in Chapter 2, the Blue Grass Chemical Agent Destruction Pilot Plant (BGCAPP) will use a number of processing steps to destroy the nerve agent-containing chemical weapons stored at the Blue Grass Army Depot, including (1) separating energetics and chemical agents from the munitions; (2) neutralizing the agents and energetics with sodium hydroxide, producing agent hydrolysates and energetics hydrolysates; (3) treating the blended agent and energetics hydrolysates in supercritical water oxidation (SCWO) reactors; and (4) treating the SCWO effluent using a reverse osmosis (RO)-based water recovery system (WRS). The ultimate end products of these processes are

- Water from the WRS of adequate quality to be recycled for reuse as quench water in the SCWO units,
- Filter cake from aluminum filtration that meets regulatory requirements for land disposal,
- WRS brines that are planned to be sent offsite for disposal, and
- Decontaminated scrap metals that can be recycled.

The focus of this chapter is on the processing steps that follow the production of hydrolysates and removal of aluminum from the energetics hydrolysate stream—namely, the SCWO system and the WRS. Because the technologies used by these processes have never before been used to process energetics and agent hydrolysates, there are uncertainties about how well the end-to-end process will perform. In particular, although SCWO technology has been used previously to mineralize organic compounds, it has not been used on the same physical and time scales as will be the case at BGCAPP—that is, 1,000 lb/hr per reactor, 24 hours per day, 7 days per week, for up to 3 years. While one of the three SCWO reactors to be used at BGCAPP has been tested extensively for periods of up to several hundred hours, the other two have not. During testing, the reactor that was tested processed hydrolysate simulants rather than the actual blends of energetics and agent hydrolysates that will be produced at BGCAPP. Consequently, the performance of the SCWO and WRS systems at BGCAPP will not be known with certainty until these systems are placed in operation with actual hydrolysate feed.

CRITERIA FOR SUCCESSFUL HYDROLYSATE TREATMENT

Two types of criteria can be used to assess the successful treatment of hydrolysate: performance requirements and performance goals. Requirements are conditions that must be met under regulatory permits and under Chemical Weapons Convention treaty obligations. Goals are primarily oriented toward process performance and schedule and are conditions that should be met to enable satisfactory system performance.

BGCAPP staff and system contractors have described their efforts to ensure that BGCAPP complies with performance requirements and satisfies its performance goals as a full-scale pilot plant using first-of-a-kind technologies such as SCWO.[1] If the criteria established for the SCWO system and the WRS cannot be met, and if this results in an increase to the munition processing time in BGCAPP, it will not be possible to achieve the risk reduction associated with destruction of the munitions stockpile in a timely manner.

Finding 6-1. The primary criteria for successful treatment of hydrolysate involve meeting regulatory and Chemical Weapons Convention requirements and meeting process performance and schedule goals for hydrolysate treatment.

The performance requirements and goals for the SCWO and WRS are based on the results of past testing, modeling, and analysis by BGCAPP and its contractors. The perfor-

[1] J. McArthur, environmental manager, BGCAPP, "Regulatory Requirements and Notifications for Offsite Shipment of Hydrolysate," presentation to the committee on January 27, 2015.

mance requirements are summarized in Box 6-1 and major performance goals established by BGCAPP are listed in Box 6-2. The operational requirements outlined in Box 6-1 generally refer to requirements articulated in the Resource Conservation and Recovery Act (RCRA) Research, Development, and Demonstration (RD&D) permit documentation and, eventually, in the RCRA Part B permit. These requirements may become more specific and/or quantitative as systemization proceeds.

The established performance goals include those outlined in Box 6-2. These performance goals are examples provided by the committee and are based on documents provided to the committee by BGCAPP (BPBGT, 2014a; BPBG, 2013; NRC, 2013). Following systemization testing to reduce performance risks, these goals may be modified to more definitively quantify successful operations.

The committee anticipates that performance goals in particular may be modified as preoperational testing of the

BOX 6-1
Requirements for the SCWO and the WRS

- Meet conditions of the RCRA RD&D Permit for GB and the RCRA Part B Permit for VX for SCWO treatment of blended hydrolysates and for downstream treatment of SCWO effluent in the WRS;
- Satisfy Chemical Weapons Convention treaty requirements for agent and munition destruction; and
- Meet or exceed requirements set by the Program Executive Officer for Assembled Chemical Weapons Alternatives regarding efficiency, cost, and schedule.

BOX 6-2
Performance Goals for the SCWO and the WRS

Effectiveness Goals

- Reliably feed agent hydrolysate, energetics hydrolysate, and blended hydrolysate to the SCWO reactors.
- Processing rates for SCWO and WRS should be compatible with agent and energetics neutralization processing rates.
- Hydrolysates should have residence time of ≥10 sec in reactors at 1,150-1,200°F and 3,400 psig to ensure acceptable destruction of all organic species.
- Each SCWO reactor should have a nominal processing rate of 1,000 lb/hr and a target availability of at least 76 percent.
- To avoid filling the hydrolysate storage tanks, SCWO availability should be no lower than 55 percent.
- There should be no more than 200 ppm aluminum in blended hydrolysate feed.
- A eutectic mixture of salts and salt additives should be monitored to ensure that salts remain molten in the SCWO reactor. Current plans are to monitor salt composition before and after additive addition to confirm addition.
- SCWO effluent should meet release specifications for total organic carbon, pH, and conductivity.
- Desired time between liner replacement should be at least 300 hr for blended GB hydrolysate and at least 400 hr for blended VX hydrolysate.
- Desired time between thermowell replacement should be at least 75 hr for all blended hydrolysates.
- RO permeate for reuse as quench water for SCWO should meet the design objective of ≤500 mg/L total dissolved solids.
- RO reject should contain about 4 wt% salts.
- Sufficient recyclable RO water should be generated to meet SCWO quench water needs.

Schedule Goals

- SCWO ramp-up should begin no later than 2 months following start of BGCAPP operations and take no longer than 2 months to achieve full processing rate.
- SCWO should destroy 921,000 gal of GB hydrolysate, 166,000 gal VX hydrolysate, and 1,380,000 gal energetics hydrolysate in an operating period of 3 years.

Cost Goals

- All operating costs, including those for preventive and corrective maintenance, infrastructure, and materials, should be within available funding.

SCWO and WRS is carried out during systemization. Also, the performance goals may be evaluated under the RCRA RD&D permit, eventually becoming permit requirements once BGCAPP transitions to a RCRA Part B permit. A failure to meet the goals listed in Box 6-2 during systemization, while potentially serious in terms of impacts on process performance, schedule, and costs, would not necessarily mean that offsite transport of hydrolysate and SCWO effluent must be carried out. However, if any failure to meet these goals becomes apparent during operations, then modifications to the equipment and procedures used could be considered. If these modifications did not improve performance, then BGCAPP might need to consider offsite transport as an alternative to onsite treatment.

Finding 6-2. BGCAPP has established a set of initial performance goals for SCWO and the WRS. These are based on past testing with simulants, simulation modeling, and analyses and, based on experience to date with the SCWO and the WRS, appear to be a realistic starting point.

Recommendation 6-1. The ability to meet the initial performance goals established by BGCAPP for SCWO and the WRS should be verified as a result of testing during systemization.

Graded Success Scale

Having multiple criteria to evaluate performance allows decision makers to document the extent of any underperformance. In other words, while a total failure of either the SCWO or the WRS is one possibility (although very unlikely), partial or temporary failures or periods of underperformance are more likely possibilities and their impacts on operations may be more difficult to define, predict, and address. For example, suppose that some of the initial performance goals are not met due to repeated equipment underperformance (e.g., a high failure rate of a high pressure pump or a higher than expected replacement rate for the SCWO liner), resulting in reduced SCWO availability and a longer operating schedule. Although these problems could most likely be addressed by equipment or operating procedure changes, they may incur increased operating costs. By having multiple performance goals, managers can take into consideration all of the impacts of actions taken to address problems and the trade-offs between meeting various criteria. Once multiple technical factors and contingency options have been evaluated using the performance criteria, the risk of system failure can be addressed. A graded evaluation of system failure risk would be useful to allow stakeholders to qualitatively rate the potential for overall program success at any point in the project. This type of graded evaluation would facilitate communication between stakeholders and allow them to track and document BGCAPP operations in a transparent and consistent way throughout the course of the project. Table 6-1, adapted from *Review Criteria for Successful Treatment of Hydrolysate at the Pueblo Chemical Agent Destruction Pilot Plant* (NRC, 2015) exemplifies a graded scale for success that could be used for evaluating SCWO and the WRS.

TABLE 6-1 Graded Success Scale for Use in Evaluating Overall Operation and Individual Treatment Processes

Grade	Definition
0	Success is practically certain (very low probability of SCWO or WRS failure): Operations are proceeding as expected. No BGCAPP actions needed.
1	High likelihood of success (low probability of SCWO or WRS failure): Actions should be taken by BGCAPP to prepare ahead of time for implementation of contingencies in the event of failures. For example, BGCAPP might begin to prepare permit modifications and planning documents.
2	Success is uncertain (moderate probability of SCWO or WRS failure): Actions should be taken to prepare for implementation of contingency operations. For example, BGCAPP might begin processing environmental documentation (permit modifications) and finalizing contingency plans, and begin to initiate changes in infrastructure to permit offsite shipment.
3	Success is unlikely with current operations (high probability of failure of the SCWO or the WRS): Actions are taken to accelerate the implementation of contingency operations and stakeholders are consulted. For example, construction of needed facilities such as new piping and loading docks is completed as quickly as possible; environmental approvals are expedited, if not already obtained; and contracts for shipment offsite and disposal at a permitted treatment, storage, and disposal facility are signed.

SOURCE: Adapted from NRC (2015).

A DECISION FRAMEWORK FOR DETERMINING SUCCESSFUL SCWO AND WRS OPERATION

Figure 6-1 presents a possible decision framework for considering the possibility of offsite hydrolysate and/or SCWO effluent treatment. It shows the high-level considerations and decision types that drive major changes to any type of operations.

This decision framework allows decision makers to evaluate all available onsite options before considering offsite shipment of hydrolysate or SCWO effluent due to underperformance or failure of the WRS. Onsite options include the risk reduction actions (i.e., testing during systemization) that will be taken to improve SCWO and WRS performance; identification of alternative water sources for SCWO should the WRS not perform; and any onsite hydrolysate processing options other than SCWO, should they exist at BGCAPP. The SCWO and WRS processes will be continuously evaluated during plant operations, with routine corrective actions taken as needed. In all likelihood, not all of the performance goals will be met all of the time, for example, there may be

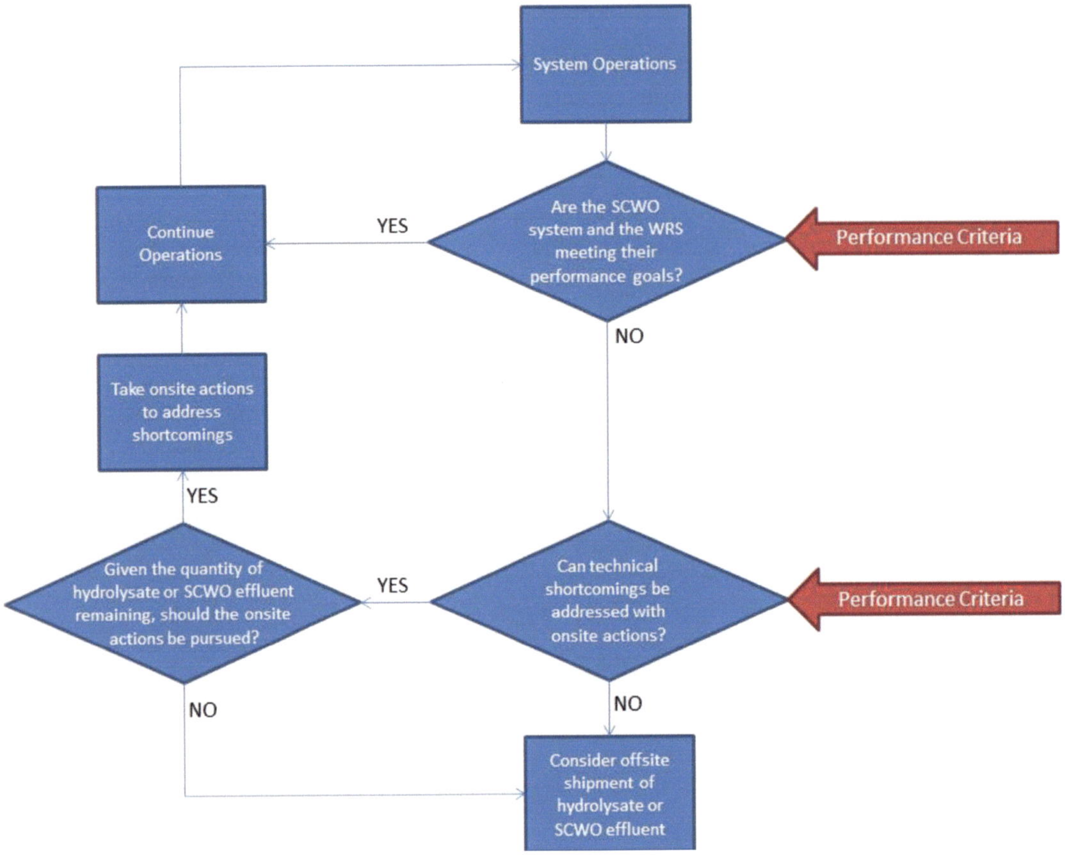

FIGURE 6-1 Decision framework for meeting performance goals.

process upsets, equipment may underperform or fail, and maintenance may take longer than expected. As a result, onsite actions may need to be taken to mitigate problems as they occur. Examples of mitigating actions include but are not limited to modifying the mix of additives to the SCWO, replacing equipment that underperforms or fails, reducing hydrolysate feed rates to the SCWO, carrying out modifications to mitigate fouling and scaling of the RO membranes in the WRS, and more frequent preventive maintenance activities. Possible mitigation actions are further discussed in Chapter 7. All of these actions would increase costs and could result in schedule delays but may be preferred to offsite shipment of hydrolysates. Other than a potential increase in risk associated with prolonged storage of munitions and hydrolysates as onsite mitigation is done, these actions should have less of an impact on BGCAPP operations, require few if any permit modifications, have little if any impact on the local stakeholders, and be consistent with the commitment to onsite treatment made by the Program Executive Office (PEO) for Assembled Chemical Weapons Alternatives (ACWA).

It should be noted that even if the SCWO or WRS underperform, substantial onsite capacity exists to store hydrolysate, allowing time for actions to be taken to improve SCWO or WRS performance while agent destruction continues. The two energetics hydrolysate tanks have a capacity of 550,000 gallons, or 40 percent of the total energetics hydrolysate production; the two GB storage tanks have a capacity of 600,000 gallons, or 65 percent of production; and the VX tank has a capacity of 80,000 gallons, or 48 percent of production.[2] Even if SCWO were to not operate at all from the beginning, it would take 36 weeks of BGCAPP hydrolysis operation before the first tanks (the energetics hydrolysate tanks) are full.[3] Also, SCWO availability, while expected to be about 76 percent, could be as low as 55 percent throughout the entire agent operations period before there would be an impact on BGCAPP agent hydrolysis operations.[4] This is predicated on the availability of quench water of suitable quality, as discussed in Chapters 2 and 7.

[2] J. Barton, BGCAPP chief scientist, Battelle, "BGCAPP Agent and Energetics Treatment Processes," presentation to the committee on January 27, 2015.

[3] BGCAPP response via e-mail to "Eight NRC Questions for BGCAPP, Rev. 2," received on February 27, 2015.

[4] Ibid.

Factors Affecting a Decision to Move to Offsite Treatment

If the schedule for overall destruction of the munitions is delayed due to the inability of SCWO to destroy all of the hydrolysate, then following consultation with stakeholders, a decision may be made to ship excess hydrolysate offsite. A decision to terminate SCWO processing and to ship hydrolysate offsite will depend on the severity of problems with SCWO, on the time during BGCAPP operations when these problems occur, and also on the degree of confidence that BGCAPP staff have in the ability of SCWO to continue to perform adequately for the remainder of its operating life, following any onsite modifications to equipment and operating procedures to mitigate problems. If the performance problems are chronic and, in the judgment of the PEO ACWA and BGCAPP staff, are not amenable to correction, confidence in SCWO may be low enough that a decision to discontinue SCWO processing might be justified. If, however, the problems are deemed to be correctable with resulting longer term improvements in performance, a decision to incur the effort of making changes may be justified. These decisions will be made during SCWO and WRS systemization and operations and, to a great extent, will be based on the judgment of the PEO ACWA and BGCAPP staff and on their continued confidence in the performance of the SCWO and WRS technologies.

Even if onsite actions are taken that will allow continued operation of the SCWO reactors and of the WRS, decisions to take these actions will be made in the context of the schedule of overall BGCAPP operations, as indicated by the lower left decision diamond in Figure 6-1. A decision to replace or repair nonperforming equipment in order to comply with hydrolysate processing goals may be economically justified early in SCWO processing, when 2-3 years of such processing remains; a similar decision may not be justifiable a month or two before the end of SCWO processing.

Any decision to consider shipping hydrolysate offsite will need to take into account potential impacts on overall BGCAPP performance and compliance with permits and treaty obligations, as well as the stakeholder perspectives described in Chapter 3. Within BGCAPP, examples of the potential impacts of shipping hydrolysate offsite include worker risks associated with hydrolysate handling, the elimination of some solid waste streams (e.g., filter cake), and making provisions for the safe and effective transfer of large quantities of hydrolysate from the storage tanks to truck loading docks, including adding new piping, road infrastructure, signage, and room for loading.[5] These impacts are discussed in greater detail in Chapter 7, in the Offsite Shipment as a Contingency Option section. The impacts will likely affect costs and the BGCAPP schedule and will need to be compared to similar impacts of making onsite changes to improve SCWO or WRS performance, if possible.

These timing considerations need to be part of the decision framework, as indicated in Figure 6-1. In the event that onsite changes cannot be made and offsite hydrolysate transport must be given serious consideration, it is prudent for longer-lead-time activities such as permit modification paperwork, preparations for making engineering and operational changes, and making contractual arrangements for offsite shipment to be carried out in advance, even if these activities will never be needed. Otherwise, the length of time needed to transition to offsite shipment, should it become necessary, could be significant.

While all stakeholders are interested in seeing BGCAPP operate successfully, not all, or even most, of these stakeholders are decision makers. The BGCAPP operations staff will be the decision makers for routine, day-to-day operations and any changes that may need to be made to these. For example, the BGCAPP operations staff would be the ones to address any problems that might be encountered in managing the composition of additives to the SCWO feed streams that are intended to control salt transport in the SCWO reactor. If an issue were to persist, the decision to pursue further alternatives might be elevated to BGCAPP management or to the PEO ACWA level and might include decisions such as more frequent replacement of the SCWO reactor liner, resulting in possible increased costs and schedule delays. Decisions having a greater impact, for example to discontinue using the SCWO reactors, would require stakeholder consultation to discuss alternatives.

PREOPERATIONAL TESTING FACTORING INTO THE DECISION PROCESS

To minimize the performance risks associated with the use of SCWO and the WRS at BGCAPP, a series of planned preoperational testing activities will be carried out during systemization. Systemization is defined as "the testing of all process components, subsystems, and systems, and the demonstration that the plant, procedures, and personnel are ready for toxic operations" (BPBGT, 2012). Systemization is divided into seven subphases (NRC, 2013):

- Presystemization
- Construction support
- Precommissioning
- Commissioning
- System startup
- System demonstration
- Optimization

The systemization process provides an opportunity to identify potential failure modes, test alternative operational and treatment strategies, and accumulate valuable SCWO and WRS operating experience prior to commencing hydro-

[5] J. Barton, BGCAPP chief scientist, Battelle, "Downstream Impacts to Plant Operations If Offsite Shipment Is Required," presentation to the committee on January 28, 2015.

lysate processing. It is expected that this testing, with its careful data collection, evaluations of system performance, and any resultant mitigation actions will reduce negative impacts to the overall operations schedule and thus also reduce overall operating costs. It will also go a long way toward avoiding a complete failure of the SCWO and WRS.

To maximize the chances of successful systemization testing, a SCWO Working Group has been established at BGCAPP to identify gaps in SCWO knowledge and experience, provide recommendations for closing the knowledge gaps, and prepare a schedule of testing and related activities intended to ensure successful SCWO start-up and operation.[6,7] These preoperational activities are intended "to address all SCWO system deficiencies identified by the [SCWO Working Group]" and will take place during the 6-month time period preceding the start of BGCAPP operations that has been allocated for the optimization subphase of systemization (BPBGT, 2014c). The activities will help to determine the likelihood that the SCWO system will meet its performance goals. They will also be used to identify and test modifications to the SCWO in case it underperforms during systemization.

The preoperational tests and related activities will address many of the technical factors that could lead to underperformance of the SCWO (BPBGT, 2014b). These activities will be carried out to address equipment issues that were identified in first-of-a-kind testing and to reduce both SCWO system risks (i.e., equipment and operational) and SCWO chemistry risks (e.g., hydrolysate feed, feed additives, aluminum removal, hydrolysate blending, and sulfur transport). Examples of the activities to be carried out include

- Evaluation of the effectiveness of sulfur blending and heating,
- Incorporate design changes from SCWO first-of-a-kind testing,
- Modification of facility control system software to reduce the incidence of unneeded shutdowns,
- Actions to improve the reliability of high-pressure air compressors,
- Testing of the integrated operation and maintenance of all three SCWO trains, and
- Review of aluminum levels in munitions and testing of reduced quench flow in the GB campaign based on aluminum levels in the hydrolysate.

These and other activities to reduce performance risk are discussed in greater detail in Chapter 7, where technical factors that could lead to SCWO and WRS underperformance are identified. For each factor, the potential impact on system performance is described, a performance grade based on experience to date is assigned, and alternative onsite contingency options are identified. Since preoperational testing to reduce performance risks will begin during the latter phases of systemization, it is premature to assign overall grades to the SCWO and the WRS.

The benefits expected from this testing include improvements in operator capabilities, safer operations, improved equipment reliability and availability, optimization of the mix of feed additives and salt transport, and reduction in materials costs (e.g., maximizing liner and thermowell life).

Finding 6-3. A preoperational SCWO testing program will be conducted during systemization. This program, developed by the BGCAPP SCWO Working Group, is intended to identify gaps in knowledge, experience, training, and equipment performance and to result in improved SCWO operations.

The committee notes that there is no WRS working group similar to the SCWO Working Group. The systemization plan for the WRS is at present very basic, with little of the detail shown in the SCWO Systemization Planning Report (BPBGT, 2014c). Examples of potential WRS issues, discussed in Chapter 7, include the likelihood of high solids loading to the multimedia filters, no apparent provision for clean backwash water, and unknown removal efficiency for the titanium dioxide particles. These types of issues will need to be examined as part of preoperational testing of the WRS units. Preoperational testing of WRS is critical to ensure that the system provides an adequate amount of quench water to the SCWO units.

Finding 6-4. No preoperational WRS testing program similar to the SCWO testing program presently exists—for example, one that describes potential performance gaps, recommendations for correcting these gaps, and a plan for implementing these recommendations.

Recommendation 6-2. BGCAPP should establish a WRS Working Group to develop a comprehensive preoperational testing program for the WRS that will identify gaps in knowledge, experience, training, and equipment performance and result in efficient WRS operations.

The preoperational test results will inform any operational changes and adjustments needed prior to hydrolysate treatment. Once actual hydrolysate processing begins, operational data will be collected to monitor performance of the SCWO and WRS against established performance criteria and to make further changes to operating procedures, process chemistry, and equipment where it is necessary and feasible to do so. If the SCWO and the WRS fail to satisfy their regulatory requirements and their performance goals despite the implementation of changes that are made following the risk

---

[6] G. Lucier, deputy chief scientist, BGCAPP, "Supercritical Water Oxidation Risk Mitigation Activities," presentation to committee on January 28, 2015.

[7] The composition of the SCWO Working Group is discussed in the SCWO Systemization and Likelihood of Insufficient Treatment section in Chapter 7.

reduction tests and despite the onsite contingency options discussed in Chapter 7, offsite hydrolysate shipment as a contingency option would need to be considered. This is also addressed in Chapter 7.

REFERENCES

BPBG (Bechtel Parsons Blue Grass). 2013. *Test Report for Supercritical Water Oxidation (SCWO) First-of-a-Kind (FOAK) Test (Preliminary Draft).* SDN-24915-00-GQY-GGEN-00094. Richmond, Ky.

BPBGT (Bechtel Parsons Blue Grass Team). 2012. Systemization Implementation Plan (SIP), Rev. 0, August 30. Richmond, Ky.: Blue Grass Chemical Agent Destruction Pilot Plant Program Office.

BPBGT. 2014a. *Characterization of Supercritical Water Oxidation (SCWO) Operating Requirements.* 24915-00-30R-SCWO-00001. Richmond, Ky.

BPBGT. 2014b. *Recommendations for Correcting Potential Gaps in Supercritical Water Oxidation (SCWO) Knowledge, Experience, and Performance.* 24915-00-30R-SCWO-00003. Richmond, Ky.

BPBGT. 2014c. *Supercritical Water Oxidation (SCWO) Working Group (SWG) Systemization Planning Report.* 24915-00-30R-SCWO-00004. Richmond, Ky.

NRC (National Research Council). 2013. *Assessment of Supercritical Water Oxidation System Testing for the Blue Grass Chemical Agent Destruction Pilot Plant.* Washington, D.C.: The National Academies Press.

NRC. 2015. *Review Criteria for Successful Treatment of Hydrolysate at the Pueblo Chemical Agent Destruction Pilot Plant.* Washington, D.C.: The National Academies Press.

7

Underperformance and Failure Risks, Systemization, and Contingency Options

Figure 7-1 provides an overview of the hydrolysate treatment system at the Blue Grass Chemical Agent Destruction Pilot Plant (BGCAPP), indicating the two major components of the system, the supercritical water oxidation (SCWO) system and the water recovery system (WRS), and the various inputs to and outputs from the overall system. This chapter discusses the possible risks of underperformance or failure of the SCWO and/or the WRS. The BGCAPP project staff is in the process of identifying sources of potential SCWO and WRS underperformance or failure and contingency options should either system underperform.[1] They plan to evaluate these risks in preoperational testing activities that will take place during systemization (see Chapter 6). This chapter also considers the risk of underperformance or failure once the system begins treating actual hydrolysate and focuses on decisions leading to possible changes in plant operations. The decision framework, the performance criteria, and the graded scale for success that were introduced in Chapter 6 are used in the discussion of technical factors in this chapter.

This chapter is organized in the following manner. For each component (SCWO and WRS), the following sections are presented:

- Technical factors that may lead to insufficient treatment,
- Systemization and likelihood of insufficient treatment,
- Impacts if the system underperforms or fails to perform,
- Contingency options for onsite operations, and
- Summary table with graded performance.

Finally, the possibility of multicomponent failure is discussed. If all contingency options are deemed ineffective, or if a multicomponent or catastrophic failure occurs, decision makers may want to consider shifting operations toward offsite shipment of hydrolysate and/or SCWO effluent. Specifically, the final section of Chapter 7 considers actions that should be taken to prepare for and implement offsite shipment of hydrolysate and the downstream impacts of this change.

UNDERPERFORMANCE AND FAILURE RISKS, SYSTEMIZATION, AND CONTINGENCY OPTIONS FOR THE SUPERCRITICAL WATER OXIDATION SYSTEM

As discussed in Chapter 2, SCWO is a suitable technology for secondary treatment of hydrolysates at BGCAPP because it can irreversibly break down a variety of organic compounds using only oxygen, water, and supplementary fuel into simple, benign molecules such as CO_2, H_2O, N_2, and inorganic salts. For example, SCWO solutions have been designed for the treatment of diverse waste materials, water purification, the recovery of precious metals from catalytic materials, and the production of heat and power. However, based on research and development of SCWO systems, including first-of-a-kind (FOAK) testing on hydrolysate simulants, BGCAPP scientists and engineers have identified a number of challenges that may be encountered in the operation of the SCWO system at BGCAPP, and they are taking appropriate corrective actions to address these challenges (BPBG, 2013). The rest of this section discusses underperformance and failure risks, and presents contingency options for use of SCWO to treat hydrolysates at BGCAPP.

Finding 7-1. BGCAPP scientists and engineers have identified a number of challenges that may be encountered in the operation of the SCWO system at BGCAPP and are taking appropriate corrective actions to address these challenges.

The feed source to the SCWO system will consist of two independent hydrolysate feed materials, as shown in Figure 7-1. One will be energetics hydrolysate combined with

[1] J. Barton, BGCAPP chief scientist, Battelle, "BGCAPP Agent and Energetics Treatment Processes," presentation to the committee on January 27, 2015.

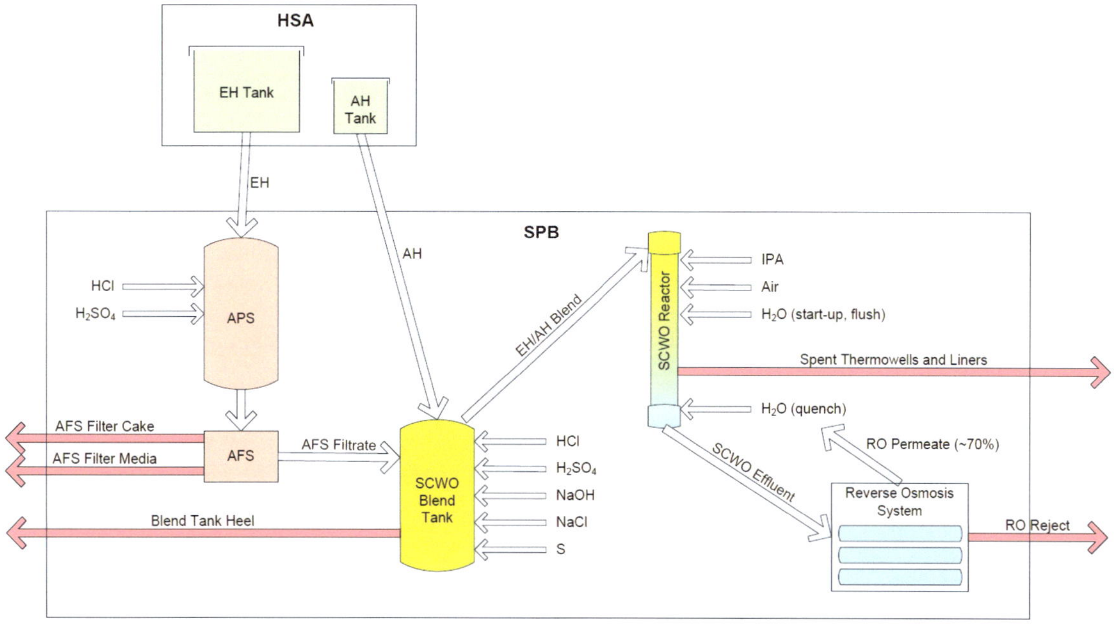

FIGURE 7-1 Schematic of the major components of hydrolysate treatment housed in the SCWO process building (SPB) including the Aluminum Precipitation System (APS) and Aluminum Filtration System (AFS), the SCWO blend tank and reactor, and the Reverse Osmosis (RO) Water Recovery System. This schematic also illustrates all of the process additives utilized and waste streams produced in the treatment. NOTE: AH, agent hydrolysates; EH, energetics hydrolysate; HSA, hydrolysate storage area; IPA, isopropyl alcohol. SOURCE: G. Lucier, deputy chief scientist, BGCAPP, D. Linkenheld, SCWO start-up supervisor, and L. Austin, waste manager, BGCAPP, "SCWO Process: Cradle to Grave," presentation to committee on January 28, 2015.

GB hydrolysate, and the other will be energetics hydrolysate combined with VX hydrolysate. While each hydrolysate feed will have similar characteristics, such as a high pH and high organic and inorganic content, the detailed composition of each material must be taken into account for successful treatment in the SCWO units to occur. Outlined below are the characteristics of each feed and the special considerations that must be dealt with in treating each component.

Energetics Hydrolysate

In addition to chemical agent, energetic materials from the chemical rockets in the stockpile will also be hydrolyzed. The motors and bursters of M55 rockets will be physically separated during the dismantling of the rocket and the energetic materials (including compound B, made of RDX and TNT, along with M28 propellant from leaker rockets) neutralized in the energetics batch hydrolyzers. The chemistry of the hydrolysis step and the resultant composition has been well documented (Masterson and Hurley, 1997; Balakrishnan et al., 2003; Newman, 1999). The unique chemistry of the energetics hydrolysis results in the potential formation of cyanide. This hydrolysis reaction has been extensively studied, and BGCAPP claims that specific hydrolysis reactor conditions have been identified that should effectively eliminate cyanide in the hydrolysate product (BGCAPP, 2013).

While the neutralization of energetic materials is effective, the resulting hydrolysate contains aluminum that is particularly problematic to the SCWO process. Past studies, using actual energentics hydrolysate, investigating the oxidation of organics with materials that contain significant amounts of aluminum have resulted in solid precipitates in the SCWO reactor that interfere with the flow of material in the reactor. As a result, the energetics hydrolysate will be treated with a precipitation and a filtration step—the aluminum precipitation system (APS) and aluminum filtration system (AFS), respectively—to remove most of the aluminum before it is transferred to the SCWO blend tanks. The precipitation is accomplished by adding acid to the energetics hydrolysate to produce an aluminum hydroxide precipitate, followed by dewatering using a filter press. The filter cake is captured in large roll-off boxes and sent for disposal offsite. The filtrate from the AFS will be collected in storage tanks, where it will be kept for blending with agent hydrolysate as feed to the SCWO.

It would be at this point in the process where, if cyanide is produced during the hydrolysis reactions, it would become particularly problematic as it would be converted to hydrogen cyanide upon acidification of the energetics hydrolysate and would be released within the headspace of the APS and AFS. Proper care must be taken to ensure that the appropriate operating conditions of the hydrolysis reaction are maintained to minimize cyanide in the hydrolysate. Proper analytical practices would need to be used to determine the presence of cyanide. If cyanide is found to be present above levels of concern, appropriate personnel protection equipment should be used to protect workers from potential exposure to cyanide. In addition, waste containing cyanide would have to be properly treated under Resource Conservation and Recovery Act (RCRA) regulations if it meets the definition of a RCRA reactive waste.

Finding 7-2. Although hydrolysis reaction conditions have been optimized to minimize cyanide production, there is a possibility that cyanide will be present in the energetics hydrolysate. Residual cyanide could become a problem in the aluminum precipitation and filtration system, where the hydrolysate is acidified, potentially producing hydrogen cyanide gas.

Recommendation 7-1. If it is determined that the energetics hydrolysate contains cyanide above levels of concern, (1) BGAPP personnel should take corrective action to optimize operational conditions so that cyanide is destroyed to acceptable levels, and (2) workers should wear the appropriate level of personnel protective equipment until such corrective action has been taken.

GB and VX Agent Hydrolysate

The nerve agent GB will be destroyed first, followed by the nerve agent VX, both using caustic hydrolysis. The chemistry of GB and VX hydrolysis has been well documented (Munro et al., 1999). After hydrolysis, the agent hydrolysates will be stored in large holding tanks. The agent hydrolysate will then be blended with energetics hydrolysate (after the removal of aluminum in the APS and AFS) as the feed mixture to the SCWO system. The blending process is very important as it has been demonstrated that there is a synergistic effect of mixing hydrolysates that optimizes performance, so the blending process needs to be properly managed.

It also needs to be noted that the GB hydrolysate can contain some dissolved aluminum due to corrosion of the GB rocket warheads (GB corrodes aluminum). The blending recipe therefore also considers the dilution needed to meet target aluminum levels in the SCWO feed that can be tolerated—that is, less than 200 mg/L.

There is one other characteristic of the GB hydrolysate that must be carefully managed. While in storage, the pH must be maintained above 12 to prevent any reformation of the hydrolysate material back to GB or other toxic byproducts (NRC, 1999). This phenomenon is well understood, and provisions are in place to eliminate this risk. Nonetheless, this operating discipline must be maintained.

Technical Factors That May Lead to Insufficient SCWO Treatment

Process flow diagrams; mass balance tables; and throughput, reliability, availability, and maintainability models have been developed by the contractor to simulate flow operations. This information provides a basis for estimates for how significantly the SCWO system can underperform and still be viable. As shown in Figure 2-1 and Figure 7-1, the SCWO system is complex and has a number of components, all of which have to operate in unison for hydrolysate to be effectively destroyed. First, agent hydrolysate and energetics hydrolysate (after aluminum removal) are mixed with specific amounts of one or more of HCl, NaOH, H_2SO_4, NaCl, and S in the blend tank. In addition to requiring specific blend ratios of energetics hydrolysate to agent hydrolysate in the SCWO feed system, salt management is also critical to the successful operation of the reactors. Each feed campaign has specific formulation requirements to generate what are characterized as eutectic mixtures of salts in the SCWO feed blend tanks. Proper maintenance of a eutectic mixture of the key salts ensures that, under SCWO operating conditions, the insoluble salt mixtures flow through the reactor system and do not precipitate inside the reactor. Even so, there are some salt components that are organically bound and thus do not become available for eutectic formation until minimum levels of oxidation occur. As a result, there is a risk of some precipitate formation in the front end of the reactor. While this is a potential risk, FOAK testing conducted during the development of the process demonstrated reasonable throughput without any major salt plugging issues. Successful operation of the SCWO during agent processing will require close monitoring of feed composition with each feed campaign to ensure the target additive concentrations are reached prior to feeding the mixtures to the SCWO reactors.

Further, while all the hydrolysates contain organic species, their heat content is insufficient to sustain the energy required for supercritical conditions. To sustain SCWO operating temperatures and pressures, a fuel (isopropyl alcohol, IPA) is co-fed to the SCWO reactors along with the hydrolysate blend. This has also been factored into the design of the SCWO and, while temperature is a critical operating parameter, FOAK testing has demonstrated that the proper operating conditions can be successfully maintained.

The mixture is then fed into one of three parallel SCWO reactors along with high-pressure air and water. After the hydrolysate has been destroyed, the reaction is quenched using recovered process water from the WRS. Gases are separated from the liquid, analyzed, and filtered. The aqueous effluent is then analyzed for total organic carbon (TOC) before being sent to the WRS. The design criterion for TOC is <10 ppm.

For the following discussions it is important to note that each SCWO reactor is fitted with a replaceable titanium liner and has titanium thermowells that contain thermocouples to monitor temperature during the SCWO. At the committee's

January 2015 meeting the following critical equipment performance points were presented. These were identified as a result of FOAK testing.[2]

Corrosion

Titanium liner and thermowell corrosion has been identified as an important limiting characteristic issue in the SCWO system (NRC, 2013). At this point, thermowell corrosion is expected to determine the maximum operating time between reactor shutdowns and maintenance. The observed corrosion rates of the liner and thermowells have been measured to be on the order of 1 mil per hour. The corrosion rate is higher for higher temperatures and flow rates (as high as 1.9 mil per hour) and lower for lower temperatures and flow rates. The corrosion damage has the appearance of erosion-corrosion, which is a phenomenon whereby the damage is much greater than would be the case for either erosion or corrosion alone. Erosion reduces the effectiveness of the protective layer that forms on corrosion-resistant materials, permitting the corrosion reaction to progress faster than expected. This may at least partially explain why the observed attack is greatest at the top of the reactor, where the flow is most turbulent and hence most erosive. Lower in the reactors, the flow pattern may be nearer to laminar flow, which is much less erosive. The presence of liquid salts on the reactor wall lower in the reactor may also contribute to the lesser damage rate by providing a physical barrier between the reactor wall and its contents. The maximum allowable extent of corrosion prior to liner replacement is 67 percent of the initial wall thickness and 80 percent for the thermowells (NRC, 2013). Thermowell and liner replacement both require shutdown of the SCWO reactor, about 6 hours for the thermowell replacement and 10-12 hours for liner replacement (NRC, 2013). It appears that the corrosion rates are predictable and that replacement can be treated as a maintenance issue rather than as an equipment failure. Nevertheless, the frequency of replacement will still have an effect on reactor throughput and will need to be monitored continuously.

High-Pressure Air Compressor Reliability

During the FOAK testing, the high-pressure air compressors were found to be unreliable and resulted in unscheduled SCWO train shutdowns (BPBG, 2013). These incidents included high-pressure cooling loop failures, expansion tank bladder failures, and low oil level/demister failures (NRC, 2013). Modifications have been made to the air compressors at BGCAPP to correct these problems. A program has been developed during systemization testing to verify that the modifications made to increase reliability of the air compressors are effective.[3] The BGCAPP SCWO Working Group (see below) also recommended manifolding the high-pressure air compressors so that all of the SCWO reactors will have a common air supply.

Compressor Condensate Oil-Water Separator Operation

The oil–water separator (OWS) is used to separate emulsified oil from the condensate discharged from the high-pressure air compressors. The oil, which comes from compressor lubrication, can account for a significant fraction of the condensate and can cause fouling of the downstream filters. While there are four high-pressure air compressors for the SCWO, there is only one OWS. During the FOAK testing, there were several instances when the OWS required workarounds to keep the system running (NRC, 2013). While OWS malfunction would probably not shut down SCWO operations, increased maintenance and repairs to the OWS system would likely reduce overall system availability and hydrolysate throughput.[4] Modifications have been made to BGCAPP's OWS (e.g., replacing the pump and cleaning the oil sensor regularly) to correct these problems. BGCAPP plans to test the modified OWS during systemization to verify that it can operate reliably. The National Research Council's 2013 report on the BGCAPP SCWO FOAK testing recommended installation of a backup OWS (NRC, 2013).

Finding 7-3. The operation of the oil–water separator (OWS) has been problematic. At present BGCAPP does not have plans to install a backup OWS.

Recommendation 7-2. BGCAPP should install a backup oil–water separator.

Liquid Effluent Letdown Valve Erosion

The liquid effluent letdown valve is located between the high-pressure gas–liquid separator (HPGLS) and the low-pressure gas–liquid separator (LPGLS) (see Figure 2-1). This valve moderates the pressure of the SCWO effluent and allows liquid and some of the gas out of the HPGLS and into the LPGLS for further separation. During FOAK testing, the Hastelloy C-276 valve seats were eroded by titanium dioxide particulates in the SCWO effluent, which originated from corrosion of the SCWO liners, resulting in excessive flow from the HPGLS to the LPGLS (BPBG, 2013). The valves, including the seats, cage, and trim, were rebuilt with Stellite 6B and will be evaluated during systemization (BPBG, 2013). This is a cobalt alloy that is harder and has better erosion resistance than Hastelloy C-276. Tungsten carbide is another alternative material being considered for the valve seat. While this, too, is harder than Hastelloy C-276, the

[2] G. Lucier, deputy chief scientist, BGCAPP, "Supercritical Water Oxidation Risk Mitigation Activities," presentation to committee on January 28, 2015.

[3] Ibid.

[4] Ibid.

tungsten carbide particles are held together by a matrix of a softer metal, such as nickel, cobalt, or other alloys, that may still be susceptible to wash-out during erosive conditions.

Solid Sulfur Heating and Mixing in the Blend Tank

Elemental sulfur is added to the blended hydrolysate to facilitate salt flow through the SCWO reactor. This sulfur will be oxidized to sulfate (SO_4^{-2}) in the reactor. However, inadequate mixing in the blend tank caused the sulfur to melt onto the heating elements during the FOAK testing, resulting in reduced and/or blocked flow through the SCWO reactor and erratic salt transport through the reactor (BPBG, 2013). During systemization, blend tank agitation and heating will be evaluated. It should be noted that the heating system in the operating unit will be different from that used in the FOAK test. Also, SCWO operation with a duplex basket strainer after the feed tank heater will be tested. Other alternatives that might overcome this problem include reducing or eliminating solid sulfur by allowing deviations from the target Cl:S ratio or by addition of an alternative nonacidic sulfur material.

SCWO Control System Code

There is a concern that the SCWO control system may produce unnecessary automated system shutdowns. The programs are being modified by BGCAPP staff to optimize the SCWO parameters and to add system holds that temporarily halt the process without causing a system shutdown (BPBG, 2013).

TOC Analyzer Reliability

Concerns have been raised that the online TOC analyzer, which samples SCWO liquid effluent to determine effectiveness of organics destruction, may become clogged by particulate material, thereby skewing the analysis results due to loss of instrument sensitivity. There are also concerns about the analyzer functioning effectively in this high salt environment. There are two TOC analyzers for each SCWO train to address this concern. The SCWO gaseous effluent is also sampled for carbon monoxide, which provides a parallel verification of organics destruction. Software and hardware modifications, including the addition of a sample extraction pump, adjusting analysis methods, and increasing sparge time along with sample acidification, will be evaluated during systemization (BPBG, 2013).

Other Concerns

Additional SCWO system concerns identified at the January 2015 meeting included these:

- Equipment deterioration during installation and preoperational storage;
- Operator currency and experience gaps;
- Unacceptable equipment wear and corrosion during operation;
- Simultaneous maintenance and operation of three reactors, which presents potential safety problems since maintenance personnel would be working near operating reactors;
- While actual agent hydrolysate and energetics hydrolysate were successfully tested on smaller SCWO reactors, only simulated hydrolysates have been tested on a full-scale unit;
- Potential variations in hydrolysate feed composition, which may present operating problems;
- The proper mix of feed additives is complex and is determined using a computer algorithm;
- Elevated aluminum concentrations in the reactor feed may cause reactor plugging; and
- Untested blend tank mixing effectiveness.[5]

BGCAPP intends to address each of these concerns during preoperational testing, concurrent with systemization.

Finding 7-4. The SCWO system is complex and has a number of components, all of which have to operate in unison for hydrolysate to be effectively destroyed in a timely manner.

SCWO Systemization and Likelihood of Insufficient Treatment

In August 2012, BGCAPP prepared a Systemization Implementation Plan for SCWO that described each element of systemization, identified the equipment and systems to be tested, and defined activities involved in plant systemization. The technical factors that may affect SCWO performance, as described above, will be addressed during preoperational testing activities, which will be carried out during the optimization subphase of systemization—that is, the 6 months immediately preceding the start of BGCAPP operations.

Furthermore, a BGCAPP SCWO Working Group was formed in April 2014 to address issues related to SCWO performance. This working group is made of representatives from

- BGCAPP engineering,
- SCWO systemization and start-up,
- BGCAPP science and technology,
- The Program Executive Office (PEO) for Assembled Chemical Weapons Alternatives (ACWA) BGCAPP field office, and
- SCWO subject-matter experts.[6]

[5] Ibid.

[6] Ibid.

The SCWO Working Group is addressing the gaps in knowledge, experience, and performance of the SCWO process, providing recommendations for closing these gaps, and producing a plan for implementing these recommendations. Its assessment and recommendations can inform BGCAPP management and stakeholders as they consider system performance and, potentially, the need for offsite shipments. The recommendations are expected to fall into four categories: administrative, maintenance, plant modification, and testing.[7] They will be addressed during systemization. With this approach, the reliability of SCWO operations should be greatly enhanced. However, in view of the complexity of the SCWO design, there is still a real possibility that unexpected issues could reduce the effectiveness and hydrolysate throughput of the SCWO process, perhaps to the point that alternative actions may need to be considered.

Finding 7-5. The SCWO Working Group is addressing concerns identified during the 2013 FOAK testing and is actively involved in planning for preoperational testing during systemization at BGCAPP.

Recommendation 7-3. The SCWO Working Group plan, as described in the December 17, 2014, Systemization Planning Report, and recommendations for correcting potential gaps in the October 27, 2014, SCWO Working Group report should be aggressively implemented. Furthermore, the SCWO Working Group should continue to provide support to all risk mitigation activities involving SCWO operations at BGCAPP.

Impacts If the SCWO System Underperforms or Does Not Perform

The underlying chemistry for destruction of the hydrolysates by SCWO is well known so that insufficient treatment of hydrolysates by SCWO is unlikely. However, SCWO operating conditions are very hard on the process systems, as illustrated by the technical factors discussed above. The two main challenges of operating SCWO process systems are salts management and monitoring and attending to corrosion (these are discussed also in Chapter 2). As a result, any system underperformance will likely be associated with an inability to keep up with maintenance requirements and equipment underperformance, causing a reduced rate of hydrolysate processing through SCWO operations. While FOAK testing established reactor downtimes and maintenance cycles, these tests were conducted on simulated hydrolysate streams and over relatively short periods of time compared to the expected 3-year destruction schedule at BGCAPP. It is, therefore, unknown whether the anticipated operational challenges—especially as they begin to compound over time—will become increasingly difficult to manage over the long term. Substantial storage capacities for hydrolysates at the front end of the secondary treatment area (discussed in Chapters 4 and 6) provide a significant buffer in the treatment timeline, affording flexibility for SCWO system maintenance and repair. However, if performance requirements and goals cannot be met consistently and satisfactorily, then it may become necessary to consider shipping some or all of the hydrolysate offsite for disposal elsewhere.

Finding 7-6. The SCWO system to be used at BGCAPP has been subjected to numerous tests with hydrolysate simulants and appears to be a mature technology capable of processing hydrolysate at BGCAPP in a timely manner. However, this technology has not been used with actual hydrolysate blends in a continuously operating environment for the 3-year time during which it is expected to perform at BGCAPP.

Finding 7-7. Although a comprehensive preoperational testing program to improve SCWO performance will be undertaken, there is still a possibility that shipping hydrolysate offsite may need to be considered at some point.

SCWO Contingency Options

Mitigation strategies for technical factors that may contribute to underperformance of the SCWO system were identified in previous reports (BPBG, 2013, and NRC, 2013). Many of these mitigation strategies and contingency options have been adopted by BGCAPP and will be tried in preoperational testing during systemization. In general, preventing equipment underperformance or failure through rigorous monitoring and maintenance of the SCWO reactors is expected to mitigate the majority, if not all, of the issues that may arise due to corrosion or salt fouling.

If corrosion is worse than expected, titanium liners and thermowells will need to be replaced on a more frequent basis. Currently, replacement of liners is anticipated every 300-400 hours and thermowells every 75-130 hours. This process requires that the reactor be taken offline while maintenance occurs. Ideally the SCWO maintenance schedule would be coordinated such that at least one of the three reactors is operating at all times, enabling uninterrupted treatment of hydrolysate. However, concerns have been raised about the safety of personnel near operating SCWO reactors (NRC, 2013). Maintenance schedules can be adjusted to address this concern.

BGCAPP is currently planning to use Grade 2 titanium in the SCWO reactors. To mitigate the challenges posed by corrosion, other titanium materials are being considered, including hardened alloys. The FOAK report suggests that they may test heat-treated Grade 2 titanium and Grade 5 titanium, which is a 6Al-4V alloy used in the aerospace industry. However Grade 2 titanium is not a heat-treatable alloy. If higher hardness from heat treatment is achieved, it will probably be because the alloy is contaminated with

[7] Ibid.

oxygen in the production process, and that is not likely to be controlled or repeatable. The corrosion resistance of Grade 5 titanium is significantly poorer than that of unalloyed titanium, such as Grade 2 titanium, but Grade 5 is harder and has greater erosion resistance than Grade 2 titanium, so if erosion is the primary cause of metal loss, there might be some improvement with Grade 5 titanium. However, if that is the case, the use of unalloyed Grade 3 or 4 titanium might be better. These grades have higher oxygen levels (Grade 2, 0.25 percent; Grade 3, 0.35 percent; and Grade 4, 0.40 percent) with associated increases in strength and hardness and no loss in corrosion resistance. The properties of these higher grade alloys would be more predictable than those of heat-treated Grade 2 titanium. However, there may be procurement issues with Grades 3 and 4 titanium. These alloys are not widely used and are probably not stocked in the warehouses of alloy distributors. They are available from alloy suppliers, but probably not off-the-shelf in large quantities, which may result in extended delivery times. If BGCAPP were to switch to alternative titanium materials in the SCWO reactors, consistent availability is an important consideration.

If there are problems with salt precipitation in the SCWO reactors, it may be due to improper levels of the sulfur additive (too high or too low), inadequate mixing, and/or heating of the solid sulfur. Allowing deviations from the target Cl:S ratio or using an alternative nonacidic sulfur material, could reduce or eliminate the need for solid sulfur. Although other strategies for dealing with salt precipitation in the SCWO reactors include periodically removing the salt deposits by brushing or scraping, or increasing operating densities to avoid precipitation altogether, these are not practical in the vertically-oriented, larger-diameter reactor vessels to be used at BGCAPP.

Finding 7-8. Due to the potential for a number of technical factors to inhibit SCWO system performance, a rigorous and efficient monitoring and maintenance program is critical to successful SCWO operations.

A number of technical factors have been identified that could potentially lead to underperformance in the SCWO system. BGCAPP has 36 weeks of hydrolysate storage capacity, which will provide some buffer to implement any needed onsite maintenance and mitigation activities. Increasing hydrolysate storage, however, will not adequately address BGCAPP scheduling challenges in the event of outright SCWO failure (as discussed in Chapters 2, 4, and 6). There are a number of reasons for this. The biggest reason is that, if BGCAPP cannot fix the SCWO or WRS in 36 weeks, the committee believes it is likely they will not be able to fix these systems at all. Further, without demonstrating that continued storage is necessary for the effective disposal of the hydrolysate, storage for longer than 1 year is prohibited under RCRA regulations. Additionally, prolonged storage would conflict with contractual requirements, would likely entail additional regulatory and environmental work and documentation, and would be problematic for certification of destruction as required by the Chemical Weapons Convention.

Technical factors leading to underperformance of the SCWO system as currently designed, along with the corresponding impacts and contingency options, are summarized in Table 7-1. Each factor is also evaluated against the performance criteria described in Chapter 6.

UNDERPERFORMANCE AND FAILURE RISKS, SYSTEMIZATION, AND CONTINGENCY OPTIONS FOR THE WATER RECOVERY SYSTEM

The WRS consists of three SCWO effluent storage tanks, a pretreatment system (ion exchange, coagulation, and filtration), three RO units (two in operation, one in reserve), two storage tanks for RO permeate, and two storage tanks for RO reject, which will eventually be shipped offsite for disposal. A brief overview of the WRS is presented in Chapter 2; a more complete description is provided in the 2012 NRC report (NRC, 2012). The WRS influent is made up of two combined streams: SCWO effluent and cooling tower/steam blowdown water.[8] These streams are high in total dissolved solids (TDS), which makes RO an attractive process for treating the water so that it can be reused as quench water for the SCWO reactors. Although RO is an established treatment technology for high TDS waters, the SCWO effluent is a unique feed for RO, which poses some treatment challenges, which are discussed below.

Technical Factors That May Lead to Insufficient Treatment

Within the WRS, the RO unit is designed to provide a 24-hour supply of recycled water for use as quench water in the SCWO reactors. In order for the RO unit to function properly, its feed must be relatively free of particulate matter and calcium, both of which may foul the membranes.

The cooling tower and steam blowdown contains dissolved calcium, which must be removed prior to RO treatment to prevent precipitation on the membranes. Currently, ion exchange is planned as a softening process for these blowdown streams prior to blending with the SCWO effluent. The efficiency of ion exchange for calcium removal, combined with the existing dissolved calcium in the SCWO effluent, would affect the ability of the RO system to meet the design recovery of 70 percent since that design target assumes no calcium in the RO influent (NRC, 2012).

Because the design recovery was calculated assuming no calcium in the influent, additional RO hydraulic design

[8] Blowdown water is water that is drained from cooling equipment or boilers to remove minerals that accumulate over time. By definition, such equipment tends to concentrate calcium and other dissolved impurities.

TABLE 7-1 Summary of Potential Technical Factors Leading to System Underperformance or Failure in the SCWO, and Corresponding Contingency Options[a]

Technical Factor	Grade[b]	Rationale for Assigned Grade	Contingency Option
Aluminum in hydrolysate	0 or 1	Excess aluminum (>200 ppm) in the SCWO feed will generate solids in SCWO system and interfere with hydrolysate processing.	• If excess aluminum in agent hydrolysate, investigate higher dilution levels to achieve target Al levels (<200 ppm). • If excess aluminum in energetics hydrolysate, adjust pH and optimize APS and AFS.
Cyanide in energetics hydrolysate	1 or 2	Excessive cyanide levels in AFS system will pose a hazard when energetics hydrolysate is acidified.	Measure cyanide levels in energetics hydrolysate; If above target, optimize energetics hydrolysis process.
Hydrolysate formulation and additives	1 to 3	If the hydrolysate composition varies significantly from batch to batch, it will be more difficult to determine the correct feed additives to prevent reactor plugging due to salt deposition and excessive corrosion.	Determine chemical composition of actual hydrolysates during operations to improve formulation strategy; storage tanks should equalize hydrolysate batches to minimize variability.
SCWO reactor liner corrosion	1 or 2	Corrosion is rapid but predictable; preoperational testing during systemization is planned to evaluate measures to increase the lifetime (i.e., protracted maintenance cycles).	• Establish regular maintenance schedule for replacement of liners to maximize SCWO availability. • Invert liners to double operating life and reduce costs. • Test other titanium materials (including hardened alloys), reduced reactor temperatures, different flow rates, and corrosion inhibitors.
Thermowell corrosion	1 or 2	Corrosion is rapid but predictable; systemization tests are planned to evaluate measures to increase thermowell lifetime.	• Establish regular maintenance schedule for replacement of thermowells to maximize SCWO availability. • Investigate larger diameter thicker-walled thermowells. • Test alternative injection nozzle velocities. • Test other titanium materials (including hardened alloys), reduced reactor temperatures, different flow rates, and corrosion inhibitors.
High-pressure air compressor reliability	2 or 3	Unreliable air compressors caused unscheduled SCWO train shutdowns during FOAK testing; modifications to the air compressors have been made and will be tested during systemization.	Manifold compressors so that they have a common air supply.
Oil–water separator reliability	0 or 1	Only one OWS in the whole SCWO system. Underperformance/failure, however, would likely not shut down the SCWO. Furthermore, the pump in the OWS has been replaced.	• Routine cleaning of the oil sensor will improve efficient separation of oil from condensate. • Install a backup OWS.
Liquid effluent letdown valve erosion	1 or 2	Hastelloy C-276 valve seats had short operating life. Valve components have been rebuilt with Stellite 6B.	Test Stellite 6B valves and evaluate alternative materials, including tungsten-carbide for valve components.
Solid sulfur heating and mixing in the blend tank	1 or 2	Inadequate mixing and sulfur melting and build-up caused reduced and/or blocked flow during FOAK testing. A different heating system and a strainer than used for FOAK testing have been added to the blend tank and will be tested during systemization.	• Improve mixing in feed tanks to keep sulfur additive in suspension. • Allow deviations from Cl:S ratio or use nonacidic sulfur additives to reduce or eliminate use of solid sulfur (this will have impacts on salt content and potentially on pH levels of SCWO feed).
Facility control system code	0 or 1	Concern over unnecessary automated system shutdown	Modify the control system to permit the control system to stop and/or temporarily halt the process without causing a system shutdown.

continued

TABLE 7-1 Continued

Technical Factor	Grade[b]	Rationale for Assigned Grade	Contingency Option
TOC analyzer reliability	0 or 1	Clogging of the TOC analyzer could cause loss of instrument sensitivity, skewing the results. Modifications have been made to both the software and the hardware. These will be tested during systemization.	• Utilize gas effluent analysis to assess hydrolysate destruction efficiency. • Have backup TOC analyzers available.
Safety issues with maintenance of SCWO reactor while other reactors are operating	1 or 2	If hazard evaluations determine that maintenance activities requiring personnel to work near operating SCWO equipment are not safe, SCWO throughput would be negatively affected.	Test integrated SCWO operational and maintenance with all three SCWO trains operating at once. Adjust maintenance schedules to reflect safety requirements.
Actual hydrolysate may behave differently from simulated hydrolysate	1 or 2	The behavior of actual hydrolysate will not be known until such materials are processed. Shifting from GB to VX hydrolysate will likely require new adjustments to operations due to different hydrolysate compositions.	Processes should be in place for adjusting additives, throughput, and maintenance schedules, as needed.
Mitigate risks from operator currency and experience gaps	0 or 1	Proactive measures are being taken to manage this risk, including training and adding experienced SCWO experts.	Ensure proper training and knowledge transfer of personnel in operations, maintenance, and analytical support divisions.

[a] Many of these contingency options are discussed in the Bechtel Parsons Blue Grass Report, *Recommendations for Correcting Potential Gaps in SCWO Knowledge, Experience, and Performance* (BPBGT, 2014).
[b] 0 = success is practically certain, 1= high likelihood of success, 2 = success is uncertain, 3 = success is unlikely.

simulations should be considered using different levels of calcium in the feed, taking into account the calcium levels in the anticipated ion exchange-treated water and SCWO effluent, in order to establish the expected level of recovery for each campaign.

Finding 7-9. While the design target for water recovery by RO was based on calculations that assumed no calcium in the RO influent, there may be low concentrations of calcium in the water collected from the cooling tower and steam blowdown, due to incomplete treatment in the water softener, or in the SCWO effluent itself.

Recommendation 7-4. Additional modeling should be conducted using different levels of calcium to verify the expected level of water recovery by RO.

In addition to having a high TDS concentration, the SCWO effluent could contain high concentrations of suspended solids. The concentrations of chemical species in the GB and VX campaigns will exceed the solubility products for $AlPO_4$, $(Ca)_2(PO_4)_3$, and Fe_2O_3 (NRC, 2012). Additionally, particulate titanium dioxide released from corrosion of the titanium liners will add to the particulate load. If these solids are not reduced to relatively low levels before the water is applied to the RO unit, their deposition at the membrane surface could lead to unacceptable pressure losses and underperformance or failure of the RO system.

Accordingly, pretreatment of water from the SCWO effluent tanks is required to remove suspended solids and to condition the water to minimize precipitation of minerals that may foul the membranes. The current pretreatment design for SCWO effluent includes coagulant addition using an in-line static mixer for coagulant dispersal and direct filtration through a multimedia filter, followed by addition of an antiscalant. One potential problem with this approach is uncertainty regarding the amounts and size of the titanium, aluminum, calcium, and iron particles and their sedimentation in the storage tanks prior to being fed to the media filters. Coagulant requirements will be dictated by the solids concentration, and this will add to the solids load. Solids overloading of the filters is a real possibility, which would lead to frequent backwashing of the filters and loss of quench water. An earlier NRC report, *The Blue Grass Chemical Agent Destruction Pilot Plant's Water Recovery System* (NRC, 2012), expressed the following concerns associated with the WRS system:

- Possible overloading of the multimedia filters with particles from the SCWO process, including titanium dioxide, iron oxides, and calcium and aluminum phosphate precipitates;
- Potential for RO fouling and scaling due to inadequate pretreatment by the coagulation and direct filtration processes; and
- Durability of the materials of construction.

Based on the briefings and documentation provided to the committee by BGCAPP, the committee believes that concerns expressed in the letter report were not completely addressed by the current WRS design.

Finding 7-10. To the committee's knowledge, the current design and systemization plan does not address many of the concerns raised in the 2012 NRC letter report *The Blue Grass Chemical Agent Destruction Pilot Plant's Water Recovery System*, including overloading of the multimedia filters with particles from the SCWO process and the potential for RO fouling and scaling.

Recommendation 7-5. BGCAPP should address the concerns raised in the 2012 NRC report *The Blue Grass Chemical Agent Destruction Pilot Plant's Water Recovery System* and incorporate the recommendations into its systemization testing and, if needed, into operational practices.

WRS Systemization and Likelihood of Insufficient Treatment

Direct filtration is a process that is not very forgiving when challenged with a high load of particulate material and the system could easily become clogged, causing short filter runs and frequent backwashing. As such, it is not typically recommended for water with high turbidity. For such water, a clarification step typically precedes filtration. Additionally, because SCWO effluent is used to backwash the filters, frequent backwashing will result in the loss of SCWO effluent to feed the RO units, increasing the need for an alternative source of water to meet SCWO reactor quenching needs. As noted in that letter report, the absence of a clarifier ahead of the filters increases the likelihood of overloading the filters with suspended solids.

Finding 7-11. The absence of a clarifier ahead of the filters increases the likelihood of overloading the filters with suspended solids.

Recommendation 7-6. The WRS design should include a clarifier before the multimedia filters to reduce the solids loading to the filters.

Uncertainties associated with solids loading and coagulant requirements for efficient multimedia filtration would ideally be addressed by preoperational testing—e.g., coagulation jar testing to determine appropriate coagulant dose and testing performance under reduced throughput—using actual SCWO effluent or a realistic simulant. These tests would also establish operational conditions for effective filtration.

It would be helpful for preoperational testing to also include testing the effectiveness of inorganic coagulants for treating SCWO effluent, and to determine the exact type(s), concentration, and operational conditions (e.g., pH) for effective coagulation. Inorganic coagulants are meant to form metal hydroxide precipitates that facilitate coagulation. However, high phosphate concentrations may lead instead to the formation of phosphate precipitates ($AlPO_4$, $FePO_4$), which could increase coagulant requirements but might not produce filterable solids. Further, it is uncertain how TiO_2 solids emanating from corrosion of the SCWO liner will respond to coagulation.

Another potential problem is the volume and quality of water available to backwash the filters. The current plan is to use SCWO effluent for this purpose. Because SCWO effluent water, as noted above, is expected to have a relatively high suspended solids concentration, the committee believes that this is not a prudent source of backwash water. Filtered water would be a better source of backwash water, but no facilities are available for storage of filtered water to meet this objective. Backwash water may be transported back to the SCWO effluent tanks for reprocessing and recovery, especially if there is a clarifier.

Finding 7-12. The source, quantity, and quality of water available for backwash of the multimedia filters are of concern. The high solids loading and use of SCWO effluent as backwash water are potential obstacles to successful operation of the WRS.

Recommendation 7-7. An appropriately sized filtered-water storage tank should be installed to provide water for backwashing the filters rather than using SCWO effluent. The WRS design should also include piping to transport backwash water to waste or for reprocessing in the WRS, according to water needs.

Finding 7-13. Since many WRS process details are unknown, including the amount of solids in the SCWO effluent, the amount of solids that settle in the SCWO effluent storage tanks, and coagulant requirements and effectiveness, successful operation of the current WRS direct filtration multimedia pretreatment system is uncertain. Therefore, successful operation of the RO units is uncertain.

Recommendation 7-8. Well-planned preoperational testing should be performed with actual SCWO effluent, or a realistic simulant, to establish operating conditions for effective pretreatment and to determine if the WRS system, especially the multimedia direct filtration system, will perform as expected. In particular, preoperational testing should determine the solids loading and corresponding coagulant requirements for effective pretreatment. As noted in Chapter 6, serious consideration should be given to forming a WRS working group analogous to the SCWO Working Group.

Impacts If WRS Underperforms or Does Not Perform

The WRS will treat effluent from the SCWO process for reuse as quench water in the SCWO units. The quench water is introduced circumferentially into the bottom of the SCWO reactor and is needed to keep salts in solution downstream of the SCWO reactor. The RO membranes should be able to provide low TDS permeate, as they are rated for a seawater influent of 3.5-3.9 percent dissolved solids content, whereas the expected SCWO effluent will have a 1-3 percent dissolved solids content. A prior NRC report found that this conservative design for the RO units should readily meet the design objectives of 500 mg/L TDS content necessary for reuse as quench water (NRC, 2012). No problems are anticipated in meeting these water quality objectives, provided that the RO pretreatment (coagulation and media filtration) meets water production needs and adequately removes suspended solids to avoid fouling and other operational problems with the RO units.

To help illustrate the linkage between WRS performance and SCWO operations, the expected flow rates for these systems are as follows: The maximum feed flow of GB or VX hydrolysate to each SCWO reactor will be 1.74 gpm, for a total of 5.2 gpm for the three SCWO reactors operating in parallel. For the GB campaign, the quench water flow demand for all three SCWO reactors is 71 gpm. In this case, the RO feed flow for the two online RO units will be 101 gpm, permeate flow will be 71 gpm (70 percent recovery), and rejectate flow will be 30 gpm. For the VX campaign, the quench water flow demand for all three SCWO reactors is 34 gpm. In this case, the RO feed flow for the two online RO units will be 49 gpm, permeate flow will be 34 gpm (70 percent recovery), and rejectate flow will be 15 gpm. In both situations, the RO permeate flow provides the entire quench water demand for the SCWO reactors. The feed flow to the RO units comprises the pretreated cooling tower and steam system blowdown flow (25 gpm for the GB campaign and 31 gpm for the VX campaign) and the pretreated SCWO effluents. The blowdown waters are necessary to meet the permeate requirements for SCWO quenching.

If the WRS underperforms or fails, not enough RO permeate will be produced to meet the demand for quench water by the SCWO reactors. The capacity of the RO permeate storage tank can meet the quench water demand for at most 1.5 days. An alternative source of clean water will need to be secured if the RO units are not able to meet the SCWO demands for quench water of 71 and 34 gpm for the GB and VX campaigns, respectively. Water from this other source might contain minerals and particulate material that could clog the SCWO nozzles, limiting SCWO operations.[9] If the quench water supply is insufficient and an alternate source of clean water is not available, then the SCWO effluent and cooling tower and steam blowdown water may need to be shipped offsite as detailed in the next section.

WRS Contingency Options

The operation of the SCWO process for destruction of the agent hydrolysates is tightly linked to the production of quench water from the WRS units. In the event of WRS underperformance or failure, such as the inability to recover sufficient quench water from the SCWO effluent, the quench water requirement would have to be rapidly filled from another acceptable water source in order to prevent a slowdown or halt in SCWO operations. For example, if the RO units are able to operate only with the pretreated cooling tower and steam blowdown water, this should yield permeate flows of 18 gpm and 22 gpm for the GB and VX campaigns, respectively. In this case, the supplemental needs for quench water would be 54 gpm for GB and 12 gpm for VX. With total failure of the WRS, the supplemental quench water needs would be the entire 71 gpm for GB and 34 gpm for VX. BGCAPP has indicated that it would likely draw from onsite surface water reserves to fulfill this need. There is currently neither infrastructure nor permits in place to make this operational change. Infrastructure or permit modifications may be necessary to do this (see Chapter 4).

If the WRS were to become nonfunctional, SCWO operations could continue if an alternative source of SCWO quench water were available. The SCWO effluent would still need to be treated or disposed of, which could involve a currently unidentified onsite process but more likely would entail offsite treatment or disposal. Here, a question would be whether to seek offsite treatment or disposal of the SCWO effluent or to forego SCWO operations entirely and send the hydrolysate offsite instead. The process flow volume does increase substantially with SCWO treatment as a result of dilution with quench water. With a total agent hydrolysate feed flow of 5.2 gpm to the three SCWO units, the SCWO effluents become 77 gpm for GB (nearly a 15-fold flow increase) and 39 gpm for VX (nearly an 8-fold flow increase). In this case BGCAPP may wish to continue SCWO operations onsite, as doing so would destroy chemical agent and other compounds regulated by the Chemical Weapons Convention (CWC), alleviating further oversight of the destruction process from the Organisation for the Prohibition of Chemical Weapons (OPCW). This is discussed also in Chapter 4. Additionally, the SCWO effluent itself, which contains, at most, 10 ppm TOC and a number of benign salt species, should not pose a significant hazard for offsite shipment.

Finding 7-14. In the event of WRS underperformance or failure (either temporary or permanent), BGCAPP may continue SCWO operations to achieve its mission of hydrolysate destruction. However, in order to do so and to ensure uninterrupted operation of the SCWO system, an alternative source of quench water would need to be established rapidly.

[9] John Barton, BGCAPP chief scientist, Battelle, e-mail communication dated February 16, 2015.

TABLE 7-2 Summary of Potential Technical Factors Leading to System Underperformance or Failure in the WRS, and Corresponding Contingency Options

Technical Factor	Grade[a]	Rationale for Assigned Grade	Contingency Option
Inadequate calcium removal during softening	1 or 2	High calcium concentrations will cause mineral scaling of RO membranes and reduce treatment effectiveness and rate of production of water for SCWO quenching	Implement monitoring of ion-exchange process for calcium; insert piping to bypass RO membranes if softening is inadequate; regenerate ion-exchange resin more frequently; implement prudent use of antiscalants.
Solids overloading of media filters due to high concentrations of metal oxides, hydroxides and phosphates, and TiO_2 released from SCWO liner	2 or 3	Solids concentration and related solution chemistry in SCWO effluent are unknown; effectiveness of inorganic coagulants is unknown; rapid headloss buildup in media filters will require frequent cleaning (backwashing), lowering rate and amount of feed water to RO units	Install clarifier to reduce solids load to filters; conduct preoperational coagulation jar tests to establish coagulant requirements.
Use of SCWO effluent for multimedia filter backwash purposes	2 or 3	Solids concentration and related solution chemistry in SCWO effluent are unknown; if solids concentration is too high or mineral deposition occurs, filters will not be cleaned to acceptable levels	Install filtered water storage tank to provide water for backwashing filters; provide associated piping needed for transporting and processing of backwash water.
Unknown water recovery efficiency	0 or 1	Current estimated 70 percent water recovery for reuse is based on simulations with no calcium content	Conduct simulations with different potential calcium concentrations to determine realistic water recovery expectations.
High phosphate concentrations could interfere with the efficacy of coagulants	0 or 1	Results in phosphate precipitates ($AlPO_4$, $FePO_4$), which will increase requirements for coagulant and may not produce filterable solids	Preoperational testing of coagulants should evaluate efficacy; pH can be altered to minimize metal phosphate precipitation; coagulant jar testing should address this concern.
Inability of WRS to meet quantity and quality of water needed for SCWO quenching purposes	2 or 3	• With the proper pretreatment facilities in place, RO system should be able to produce the water needed for quenching in SCWO operations • If pretreatment and RO production cannot keep pace with SCWO quenching needs, an alternative source of water will be needed	Draw upon supplemental local Blue Grass Army Depot plant water to provide quench water for SCWO system; arrange for addition of the necessary infrastructure to provide local water of acceptable quality for SCWO quenching; and acquire the appropriate permits to use local water for this purpose (see Chapter 4).

[a] 0 = success is practically certain, 1= high likelihood of success, 2 = success is uncertain, 3 = success is unlikely.

Recommendation 7-9. BGCAPP should implement a backup plan to provide quench water for the SCWO reactors in the event of WRS underperformance or failure, including adding the necessary infrastructure for an alternative water source, assuring that the quality of that source is acceptable for SCWO quenching needs, and acquiring the appropriate permits.

Technical factors leading to underperformance of the WRS, as currently designed, along with the impacts and contingency options, are summarized in Table 7-2. Each factor is also evaluated against the performance criteria described in Chapter 6. The lower the grade, the lower the probability of underperformance or failure.

Tables 7-1 and 7-2 list factors that could contribute to underperformance or failure in the SCWO system and the WRS. The committee assigned individual grades to each technical factor in these tables using the guidelines described in Table 6-1. Decision makers could use a similar process to evaluate the state of operations over the course of actual hydrolysate treatment. Ideally, these types of grades could be assigned to the individual technical factors, as the committee has done here for each of the subsystems, SCWO and WRS, and even to the overall hydrolysate treatment system (SCWO and WRS together). While the committee is comfortable assigning grades to the individual technical factors discussed in this chapter, it is reluctant to assign overall grades to the SCWO or the WRS at this point because BGCAPP is still 3 years away from operation and also because preoperational testing during systemization will help to alleviate concerns. In the next 3 years, many aspects of the BGCAPP operating plan will evolve, hopefully improving the likelihood of success.

OFFSITE SHIPMENT AS A CONTINGENCY OPTION

In the event of SCWO system and/or WRS underperformance, decision makers would likely first consider onsite actions that can be taken to address the shortcomings in the performance. This chapter has discussed a number of potential modifications to the hydrolysate treatment pro-

TABLE 7-3 Combined SCWO and WRS Failure Scenarios

SCWO	WRS	Contingency Option
Functions	Functions	Continue operations as planned.
Functions	Fails to treat SCWO effluent, but functions for cooling tower steam/ blowdown water	*Option 1* • Continue to process hydrolysate with SCWO, but ship SCWO effluent offsite, and • Continue processing blowdown in WRS for use as quench water in SCWO, and • Supplement SCWO quench water with demineralized Blue Grass Army Depot water as needed. *Option 2* • Halt both SCWO and WRS operations and send hydrolysate offsite.
Functions	Fails to treat both SCWO effluent and cooling tower steam/blowdown water	*Option 1* • Continue to process hydrolysate with SCWO, but ship SCWO effluent and blowdown offsite for disposal, and • Use demineralized Blue Grass Army Depot water for SCWO quench water. *Option 2* • Halt both SCWO and WRS operations and send hydrolysate offsite.
Fails	Functions	Halt both SCWO and WRS operations and send hydrolysate offsite.
Fails	Fails	Halt both SCWO and WRS operations and send hydrolysate offsite.

cesses (SCWO and WRS) in detail, weighing the causes and impacts of underperformance or failure against the onsite contingency options. However, there may be scenarios in which SCWO system underperformance is so severe, compounded, or chronic that onsite mitigation actions are no longer sufficient. In this case, offsite shipment of hydrolysate will need to be considered (see the decision framework in Chapter 6, Figure 6-1, and the discussion of factors entering into a decision to ship hydrolysate offsite, also in Chapter 6). On the other hand, if the SCWO system performs adequately but the WRS underperforms or fails, there may be workarounds that enable BGCAPP to continue the destruction of hydrolysates using SCWO. The SCWO versus WRS scenarios are summarized in Table 7-3.

Were SCWO to function properly and the WRS to underperform, BGCAPP would have an option to continue SCWO operations and send SCWO effluent offsite or, alternatively, halt both SCWO and WRS operations and send hydrolysate offsite. However, to address CWC requirements and alleviate any requirements for further OPCW oversight, BGCAPP could continue SCWO operations and send SCWO effluent offsite. As discussed in Chapter 5, on transportation, however, sending SCWO effluent offsite would substantially increase the number of offsite waste shipments. The resource requirements for increased offsite shipment versus the alleviation of requirements for further OPCW oversight associated with continuing SCWO operations would need to be carefully evaluated.

Although the offsite shipment and disposal of hydrolysates are viewed as a last resort, findings and recommendations presented in Chapters 3, 4, 5, and 6 indicate that this scenario needs planning and preparation to avoid delays in chemical agent destruction should offsite shipment prove necessary. This committee was tasked to evaluate offsite shipment of hydrolysates, not such shipment of other potential waste streams (e.g., SCWO effluent), but, as Table 7-3 indicates, there may be a need to consider offsite shipment of SCWO effluent if the WRS fails to perform adequately and contingency plans need to be developed. In view of the committee's statement of task (see Appendix A), the remainder of this chapter considers major downstream impacts of the decision to ship and dispose of hydrolysates offsite and the actions that need to be taken to achieve this objective.

As discussed in Chapter 6, the committee recognizes that there may be scenarios where onsite mitigation actions are available to remediate underperformance, but given the timeline of treatment and the quantity of hydrolysate remaining, offsite shipment may become the preferred option for decision makers and stakeholders. Along these lines, if the SCWO system or WRS are down for an extended period of time, decision makers may also evaluate options for temporary or supplemental (i.e., parallel) offsite shipment of hydrolysate with the intention of resuming full treatment capacity onsite after the secondary treatment systems are repaired. The feasibility and value of these options similarly depends on where BGCAPP is in the treatment schedule for hydrolysates, as discussed in Chapter 6.

The committee, PEO ACWA, and BGCAPP discussed the downstream impacts of offsite shipment of hydrolysate at their January 2015 meeting.[10] Many factors highlighted

[10] J. Barton, BGCAPP chief scientist, Battelle, "Downstream Impacts to Plant Operations If Offsite Shipment Is Required," presentation to the committee on January 28, 2015.

in this section were addressed at that meeting. In general, the downstream impacts can be categorized in terms of their effect on plant (physical infrastructure), paper (e.g. regulatory permit modifications), and people (e.g. BGCAPP staff, local community members, PEO ACWA)—a categorization scheme used by PEO ACWA. Although a switch to offsite shipment, if initiated, will require a significant investment of time, labor, and funding, there may also be some benefits that would be gained from the change. These benefits are indicated as well.

Plant

The hydrolysates (GB, VX, and energetics) each have unique chemical compositions that must be considered when making operational changes, including offsite shipment. In the current operating plan, the energetics hydrolysate is treated to remove aluminum and then blended with nerve agent hydrolysate and spent decontamination solution prior to treatment with SCWO. If offsite shipment of hydrolysates is pursued, BGCAPP has indicated that the three hydrolysate wastes would be shipped separately rather than blended and shipped as a mixture. If this is the case, waste characterization data would need to be provided to the receiving treatment, storage, and disposal facility or facilities (TSDFs) for each of these waste streams to inform the TSDF(s) of the waste's composition (see Recommendation 4-2). Along these lines, the committee believes it would not be necessary to remove aluminum from the energetics hydrolysate prior to offsite shipment, as long as the receiving TSDF is able to accept the hydrolysate as is. It would also be equally acceptable to ship the hydrolysates separately or blended.

Finding 7-15. If offsite shipment of hydrolysate is necessary, there is no reason to remove the aluminum from the energetics hydrolysate prior to shipment.

Should the decision be made to ship hydrolysate or SCWO effluent offsite, additional infrastructure would be needed to efficiently and effectively transfer the material for shipment. BGCAPP plans to ship the RO reject effluent and aluminum filter cake for disposal offsite. As a result, some infrastructure is already in place for truck-based shipment of hazardous waste. As discussed in Chapter 5, switching to offsite shipment of the hydrolysates prior to any additional treatment (dilution, removal of aluminum, and treatment with SCWO) actually produces fewer truckloads of hazardous waste material requiring disposal than the current plan. There is also a railway near the Blue Grass Army Depot that could be adapted for shipment of hydrolysates. Necessary additional infrastructure would likely be minimal but might include additional piping, leak and odor containment, agent monitors, waste loading areas, truck loading docks, new rail/roadways onsite, new signage, and extra traffic controls at BGCAPP.

Besides the reduced volume of hazardous waste requiring offsite shipment, there would be additional benefits to not operating the SCWO treatment system. These include elimination of solid waste streams (aluminum filter cake, spent carbon), a reduction in utility loading (e.g., power, cooling water, steam) and chemical usage, and a reduction in emissions. There would also be a significant reduction in maintenance and manpower needs, which could be seen as a benefit to some and a detriment to others. The SCWO and WRS processes would cease to operate if the offsite shipment scenario is implemented. Discussions with PEO ACWA and local stakeholders, including the Kentucky Department for Environmental Protection, would need to be held to gather input for decisions and procedures to be implemented in the event of offsite shipment of hydrolysate.

Paper

If a decision were made to ship hydrolysate or SCWO effluent offsite, a significant amount of paperwork would be required for permit modifications, changes to operating procedures, and execution of TSDF and shipping contracts that must be in place before actions can be taken to implement the decision. Chapter 4 discusses permit requirements applying to changes in BGCAPP operations and offsite shipment of hydrolysate or SCWO effluent in detail. Although the BGCAPP RCRA Research, Development, and Demonstration (RD&D) permit does not include offsite shipment of hydrolysate, offsite shipment is not prohibited by regulation. A decision to halt SCWO operations and ship hydrolysate, however, would negate the RD&D permit because SCWO was used as the basis for the FOAK technology determination underlying the RD&D permit. Offsite shipment would require substantial permit modifications, which, as discussed in Chapter 4, the committee believes would take more than a year to accomplish.

In addition to RCRA permit modification, offsite shipment might impact National Environmental Policy Act requirements and could also impact OPCW treaty monitoring requirements. BGCAPP would also need to revise the Madison County host community certification if infrastructure improvements require changes in the Emergency Response Plan. Finally, the water withdrawal permit from Lake Vega on the Blue Grass Army Depot may also need to be amended. These are also discussed in Chapter 4.

Furthermore, BGCAPP will need to identify and place contracts with licensed hauler(s) and TSDF(s), plus coordinate shipment to TSDF(s) (see Chapters 3 and 5). Changes to the BGCAPP facility's Site Safety Submission Document would be required, along with revisions, cancellations, and the adoption of new standard operating procedures and destruction schedules. "Paper" is an area where preplanning for this last-resort contingency option would be very beneficial.

People

As indicated throughout this report, there are many stakeholders in the BGCAPP project. A major change in operations, such as a shift to offsite shipment of hydrolysate rather than using the SCWO and WRS, would have a major impact across all stakeholder groups. In particular, shutdown of the SCWO would result in BGCAPP staff reductions or reassignments, and delays in implementation of offsite shipment could result in further loss of staff from the facility as a whole. The surrounding community, represented by the Citizens' Advisory Committee and the Chemical Demilitarization Citizens Advisory Board, would also be impacted by this decision. A decision to ship hydrolysates offsite would create new stakeholders, including, potentially, regulators in recipient states and residents of communities near the disposal location.

Finding 7-16. A decision to ship hydrolysate offsite could have serious impacts on stakeholders, BGCAPP operations, regulatory compliance, and obligatory requirements under the Chemical Weapons Convention. There might be additional negative impacts if BGCAPP is not prepared ahead of time for a possible transition to offsite shipment, if and when such a decision is made.

A decision to ship SCWO effluent offsite would also affect stakeholders. However, because the SCWO effluent is a nonhazardous dilute brine, the impact on the recipient community would be minimal except for increased truck traffic.

Treatment Timeline in the Offsite Shipment Decision

Any decision to ship offsite needs to be based on a set of agreed-upon performance criteria, as discussed in Chapter 6. However, an important factor that must be considered in the decision is the point at which it takes place in the treatment schedule—that is, how much hydrolysate has been treated and how much still remains to be treated (as discussed in Chapter 6). For example, if 90 percent of the hydrolysate has been treated if and when the SCWO operation or the WRS fails to perform adequately, it may be desirable to ship the remaining 10 percent offsite rather than repair the system. In contrast, if only 5 percent of the treatment has been completed when there is serious underperformance or a failure in the SCWO system or the WRS, decision makers may consider it worthwhile to take corrective actions to allow continuing onsite treatment despite possible costs and schedule delays. This is a fairly simple example; in reality the decision would likely be much more complex. As noted in Chapter 6, the decision should take into consideration the history of SCWO and/or the WRS operating problems, the severity of these problems, the costs and time needed to restore SCWO and/or WRS to its desired performance level, and, in general, the degree of confidence that the BGCAPP staff has in the ability of SCWO and the WRS to operate successfully if changes were made to equipment and operational procedures.

Among the many factors to be considered, the storage capacity for the various hydrolysates is a major factor impacting the treatment schedule. The large amount of storage capacity at BGCAPP affords a significant buffer in the schedule—at least 36 weeks—for repair of the SCWO or WRS without halting the front end neutralization processes (storage capacity is discussed in detail in Chapters 2, 4, and 6). While this creates a bias in favor of repairing the SCWO system or WRS, if repair is feasible, schedule progress, including buffer capacity, would need to be taken into account in the decision to consider shipping offsite.

The committee also deliberated at length on whether the decision to ship offsite should be permanent, or if there are scenarios in which offsite shipment could be used temporarily while modifications to SCWO and/or the WRS are made, or used in parallel while the SCWO system and WRS operate at reduced availability. Implementing offsite transport of hydrolysate under any circumstance (temporary, parallel, or permanent) will affect plant, paper and people as discussed above. Physical changes to the plant, changes in permit documentation and standard operating procedures, transportation risk assessments, retraining and the possible reassignment or even furlough or layoff of staff would need to be considered, to name just a few examples. Any effort to implement offsite transport will be considerable. Likewise, the effort to shift back to onsite treatment after a delay resulting from repairs would also be substantial.

Offsite shipment operating in parallel with reduced onsite treatment might alleviate some of the transition burdens, but the scenarios in which this option would be practicable are limited. Also, operating SCWO at reduced capacity while also shipping some hydrolysate offsite would increase the system management efforts, in addition to whatever efforts are required to repair the underperforming component(s). The committee acknowledges that at this time it is not possible to predict the exact circumstances of a SCWO or WRS underperformance or failure once BGCAPP enters into operation and that the evaluation of whether to ship offsite permanently, temporarily, or in a parallel manner is more appropriately made by decision makers and stakeholders when the specific circumstances are known. Thus, the committee makes no specific recommendations concerning the nature, extent, or duration of any option for offsite shipment of hydrolysate.

Finding 7-17. Planning and implementation of offsite shipment of hydrolysate on a temporary or parallel basis will require the same effort as will permanent offsite shipment.

Throughout the report, the committee has recommended that BGCAPP take actions, such as filing the necessary permit modifications and installing shipping infrastructure, to prepare for the last-resort scenario to ship hydrolysate offsite in order to avoid further delay in munitions processing. However, the committee also recognizes the tension that this creates in the decision process by implementing these measures before they may be needed. Making these preparations beforehand should in no way bias the decision in favor of offsite shipment. In the event that offsite shipment must be considered, the decision must be based on the application of an established decision framework and appropriate consultation with stakeholders.

Finding 7-18. The SCWO system to be used at BGCAPP has been subjected to numerous tests with hydrolysate simulants and appears to be a mature technology. Likewise, the RO system at the heart of the WRS is a proven technology for desalinating water. However, these technologies have not been used with actual hydrolysates in a continuously operating environment for the 3 years during which it is expected to perform at BGCAPP.

Recommendation 7-10. Although the SCWO and WRS appear to be capable of processing hydrolysate at BGCAPP, and a comprehensive preoperational testing program to improve performance will be undertaken, there is still a reasonable possibility that at some point during BGCAPP operations, a decision may need to be made to ship hydrolysate or SCWO effluent offsite. As a precaution, BGCAPP management should prepare for this contingency by taking all necessary actions having long lead times well in advance of such a decision.

REFERENCES

Balakrishnan, V.K., A. Halasz, and J. Hawari. 2003. Alkaline hydrolysis of the cyclic nitramine explosives RDX, HMX, and CL-20: New insights into degradation pathways obtained by the observation of novel intermediates. *Environmental Science & Technology* 37(9): 1838-1843.

BGCAPP 2013. *Energetics Neutralization–Mitigation of Low-Level Cyanide Hazard*. 24915-00-30H-G01-00133, Rev 0, WP155, July 30.

BPBG (Bechtel Parsons Blue Grass). 2013. *Test Report for Supercritical Water Oxidation (SCWO) First-of-a-Kind (FOAK) Test (Preliminary Draft)*. SDN-24915-00-GQY-GGEN-00094. Richmond, Ky.

BPBGT (Bechtel Parsons Blue Grass Team). 2014. *Recommendations for Correcting Potential Gaps in Supercritical Water Oxidation (SCWO) Knowledge, Experience, and Performance*. 24915-00-30R-SCWO-00003. Richmond, Ky.

Masterson, W.L., and E.J. Slowinski. 1972. Chemical Principles. Philadelphia, Pa.: W.B. Saunders & Co.

Munro, N.B., S.S. Talmage, G.D. Griffin, L.C. Waters, A.P. Watson, J.F. King, and V. Hauschild. 1999. The sources, fate, and toxicity of chemical warfare agent degradation products. *Environmental Health Perspectives* 107(12): 933-974.

NRC (National Research Council). 1999. *Review and Evaluation of Alternative Technologies for Demilitarization of Assembled Chemical Weapons*. Washington, D.C.: National Academy Press.

NRC. 2012. *Letter Report: The Blue Grass Chemical Agent Destruction Pilot Plant's Water Recovery System*. Washington, D.C.: The National Academies Press.

NRC. 2013. *Assessment of Supercritical Water Oxidation System Testing for the Blue Grass Chemical Agent Destruction Pilot Plant*. Washington, D.C.: The National Academies Press.

Newman, K.E. 1999. *A Review of Alkaline Hydrolysis of Energetic Materials: Is It Applicable to Demilitarization of Ordnance?*, IHTR 2167. Indian Head, Md.: Naval Surface Warfare Center.

Appendixes

A

Chronology of Events at the Blue Grass Army Depot to the Present, With Focus on Public Involvement

The process of developing and implementing the plan for destroying the nerve agent-containing munitions at the Blue Grass Chemical Agent Destruction Pilot Plant (BGCAPP) has a complex history. It informs the way that the Program Executive Office (PEO) for Assembled Chemical Weapons Alternatives (ACWA) and stakeholders might now approach consideration of a backup plan in the event that onsite treatment processes fail to perform as expected. Aspects of this history that are critical to informing discussions of a backup plan are

- The emergence of "critical trust" after the severe erosion of trust between the community and the Army;
- An expectation that community members play a meaningful role in decision making about facility design, monitoring, and performance assessment; and
- Concerns about the impacts of offsite shipments of secondary wastes such as hydrolysate.

This appendix summarizes the history. It provides an overview of the chronology and events that inform existing relationships and describes the perspectives of stakeholders and decision makers.

EARLY TENSIONS AND THE EROSION OF TRUST IN THE CHEMICAL WEAPONS DESTRUCTION PROGRAM

The initial proposal by the Army was to destroy the U.S. stockpile of chemical weapons and agents using incineration technology. Incineration was viewed as the best and most expedient option and was used widely to treat and destroy hazardous wastes in the 1980s. Incineration was also controversial, with many communities opposing the siting and operation of incinerators for the processing of hazardous and municipal wastes. In this context, it is no surprise that the early history of the chemical weapons destruction program was characterized by significant debate about the Army's preference for incineration to dispose of the stockpile. Much of this debate had its origins in questions raised by residents living nearby the Blue Grass Army Depot (BGAD).

When the Army held public meetings in the communities with facilities housing chemical weapons, the meeting in Madison County, Kentucky, had the most people attending. Rather than building support for the Army's plans to incinerate the weapons, this meeting "sparked fears in local residents, who began to organize against incineration" (Futrell and Futrell, 2012, p. 171). Some nearby residents did not know that chemical weapons were stored in their community. They also felt marginalized and disrespected by the Army, which lacked answers to their questions. Two participants described the initial meetings with Army this way:

> Going into that meeting…a lot of us were skeptical about the plan but that didn't mean we were necessarily opposed to it. But they came in, treated us like children, as if we didn't know a thing. Then that little lady asked the simplest question—What's left over after the weapons are burnt? They couldn't answer. That's about the time when we started thinking a little harder that this might be a really bad idea. (Futrell and Futrell, 2012, p. 174)

> They [the army] would waltz into every public meeting with their cadre of engineers and staff acting like we were stupid hayseeds with no chance of understanding them. We were the ones asking questions that they couldn't or wouldn't answer with anything more than a "don't worry about the details; we'll take care of them." They couldn't tell us why neutralization wouldn't work. They couldn't tell us why we should be confident in them to solve the problems they began having with incineration. They were like a parent scolding a kid who keeps asking why something is the way it is saying "because I said so." That kind of answer is just not acceptable. (Futrell, 2003a)

As residents began to ask more questions during the late 1980s through the 1990s, the process became increasingly adversarial (Futrell 2003a; Futrell 2003b) and distrust in the

Army and its plan for incineration grew. Public involvement by the Army was described as perfunctory, such that "locals soon concluded that the Army was only going through the motions of public involvement to meet their legal obligations, with no real intent to meaningfully involve them" (Futrell and Futrell, 2012, p. 175). Further, observers claimed that

> Information was highly controlled, and army interaction with citizens was mechanistic and dismissive, reflecting the common top-down model of communication in which technical experts dominated decisions. The result was hollow proceduralism, tokenistic consultation, and conflict as citizens sought a meaningful role in the process. (Futrell, 2003a, p. 460)

Public opposition was reflected in political opposition. Federal, state, and local government officials began echoing public concerns and acting on them (Futrell, 2003b).

BUILDING A NETWORK OF OPPOSITION

The community members opposing the Army's plan for incineration of the weapons at BGAD organized themselves into, first, the Kentucky Environmental Foundation (KEF), and, later, the Chemical Weapons Working Group (CWWG). The CWWG, established in 1991, is a coalition of community members from the nine chemical weapons sites around the United States, and from national and international nongovernmental organizations. Through a process of learning and dialogue, activists in the nine states began to develop a more unified view about how to handle the stockpile. Initial reactions from opponents to incineration were to ship weapons and wastes elsewhere for treatment. This preference soon bumped up against activists and residents in other locations, such as Utah, which viewed such an approach as ethically suspect by imposing new risks on additional communities (Marshall, 1996; Futrell, 2003b).

The CWWG adopted principles of environmental justice as a foundation for its proposals (Marshall, 1996). The result was a more systems view within the CWWG and a strategic shift toward supporting onsite closed-loop destruction and treatment of wastes.

> [Alternative technologies] was a wholesale shift in our thinking. We went from advocating shipping this stuff to Utah to realizing that, not just politically but ethically, that just wouldn't work. We don't want it; why should they? Perfectly logical—it just wasn't a big part of our thinking before we began talking. Considering this thing from their side was a real eye-opener and it fundamentally changed what we were trying to do. (Participant quoted in Futrell, 2003b)

Members of the KEF and CWWG, some of whom continue to be active today on Citizens' Advisory Commissions (CACs) at Pueblo and Blue Grass and include the co-chair of the Chemical Destruction Citizens' Advisory Board (CDCAB), developed considerable expertise about these technologies and their site. Futrell and Futrell (2012, p.182) describe their expertise this way:

> The most persistent KEF/CWWG activists have been involved in the dispute for more than twenty years. Through their unique and invaluable experiences, they acquired a blend of knowledge and communication skills that only they possess. Many parties in the dispute, including legislators, regulators, and even army officials, have great regard for their brand of expertise and draw upon it frequently. . . . We describe this expertise as a combination of several interconnected dimensions: holistic knowledge, issue memory, vernacular translation, and the ability to cultivate alliances.

A SHIFTING DYNAMIC

A dramatic shift in the dynamic of growing social distrust and Army insistence on incineration occurred in 1997. A number of factors led to this shift, including mounting pressure on the Army from Congress and in local communities about the safety of incineration in general and the facilities in the Army specifically, as well as the transferring of risks to new communities by transporting and disposing of weapons and wastes (Durant, 2007; Futrell and Futrell, 2012; GAO, 1995a, 1995b). This pressure emerged in large part through the efforts of the KEF and CWWG. Congressional leaders were also concerned about the prospect of interstate transportation of weapons and waste (Durant, 2007). Public Laws 104-201 and 104-208 froze funds for construction of chemical agent destruction facilities at Blue Grass and Pueblo and directed the Army to demonstrate at least two alternatives to incineration for the destruction of the agent. What would eventually become the PEO ACWA was established to evaluate other means of destroying the chemical agent.[1]

As part of the PEO ACWA history, Congress also required a more robust effort to involve the public. PEO ACWA was, in large part, the result of an attempt to address public distrust of the Army that appeared to many to insist on pursuing incineration without justification. One of the first acts of what would become PEO ACWA was to initiate the Dialogue on Assembled Chemical Weapons Assessment (ACWA Dialogue) in 1997 (Goldberg, 2003; Keystone, 2004). The ACWA Dialogue was facilitated by the Keystone Policy Center (formerly the Keystone Center), a nongovernmental organization dedicated to supporting the resolution of policy conflicts by inspiring "leaders to rise above entrenched positions to reach common higher ground."[2] The ACWA Dialogue included 32 participants from the affected communities and states and met 13 times over 5 years. The goal

[1] When first established, ACWA was the Assembled Chemical Weapons Assessment program. It then became the Assembled Chemical Weapons Alternatives program, and then PEO ACWA.

[2] More information about the Keystone Policy Center is available at www.keystone.org.

of the ACWA Dialogue was to "ensure a marriage of the best science available while incorporating the concerns of the communities and the political realities of the disposal issue" (Futrell, 2003a). Input was sought from the participants about technical and social criteria for comparing alternative technologies, assessment of the alternative technologies using the criteria, and identification of sites appropriate for the implementation of the alternative technologies. In addition, a subset of ACWA Dialogue members participated in the ACWA Technical Evaluation Team. These members and a consultant made up a Citizens' Advisory Technical Team (CATT) "that directly monitored the procurement process and also participated in the evaluation of the demonstration test results" (Goldberg, 2003, p. 317). The CATT was a critical part of the process, allowing Dialogue participants access to closed, technical deliberations where technologies were ranked. CATT members had to sign confidentiality agreements. While the CATT members did not participate in the ranking of alternative technologies, they were able to observe that the criteria and weightings developed as part of the Dialogue were used as intended. As the current chair of the Blue Grass CAC observed, "CATT was critical. We were observers, we were behind the scenes to make sure it was all above board." He further elaborated his view:

> We could not know if the criteria would be legitimately applied to the technologies because it would be in a closed room. We felt it was not legitimate—we wanted to make sure that criteria were used as we proposed, not let them tweak. So, the government selected 4 [Dialogue participants] to observe. They could not come back and explain exactly what happened but [had] to come back and present their impressions of the legitimacy of the process that occurred, so when we narrowed the options we would have full confidence about that process. I don't know if that was ever done before. But it was a remarkable accomplishment.[3]

Goldberg (2003) concludes that "rather than seeing the [DOD] as lacking interest in the effects of its actions, the ACWA Program was seen as a cooperative effort between the public and the government" and furthermore, that the public's input "was valuable and their existence increased the level of public acceptance over the course of the program" (Goldberg, 2003, p. 319; Futrell, 2003a). This was not just the view of the Army, but also of community members and congressional staff (Futrell, 2003a).

Published accounts of the ACWA Dialogue highlight its importance for developing public support, enhancing PEO ACWA responsiveness to public concerns and preferences, building the capacity of local stakeholders to participate in highly technical discussions, and rebuilding trust (Goldberg, 2003; Futrell, 2003a; Futrell and Futrell, 2012). For example, Keystone staff observed that "Although there remained strong and differing opinions at the end of the Dialogue process, there was significantly greater understanding and trust of parties on almost all sides" (Keystone, 2004). Describing the experience with the ACWA Dialogue and the prior history, a participant said

> There's a real sense of back and forth trust developing, almost trust anyway, at least starting to believe each other, work with each other, and that development is really, really different from what's been going on in the last decade. The army versus us thing that has been really characteristic just isn't there with this. There is certainly a degree of caution. I think it's on both sides. And I think it's more of a caution of "we don't want to screw this up." Everyone is very concerned about building a kind of relationship and atmosphere in which we can do things together. It's from a change in personnel; it's from a change in process; it's from a change in attitude. There's a whole new dynamic going on that is really different. (Futrell, 2003a)

Subsequently, PEO ACWA managers worked hard to build on these experiences and continue developing trust and collaboration into the PEO ACWA program. For example, locally staffed CACs and outreach programs, which play an active role in community life, were established at the sites.

The significance of the ACWA Dialogue to subsequent planning activities rests in:

- The Army's better understanding of community concerns and preferences;
- Community stakeholders feeling heard, by recognizing their input reflected in Army decisions;
- A rebuilding of collaborative relationships that formed the basis for continued deliberations and planning for BGCAPP;
- An expectation by members of the public for continued public involvement in all aspects of planning, decision making, and monitoring;
- PEO ACWA's commitment to meaningful public involvement;
- Building the capacity of local stakeholders to participate in highly technical discussions about the alternative technologies, which at BGCAPP include the supercritical water oxidation process; and
- Public acceptance of plans for destroying the stockpile.

PUBLIC INVOLVEMENT AT BGCAPP

After the technologies had been selected, the ACWA Dialogue Group was disbanded in favor of CACs, which were based in Colorado and Kentucky. The Kentucky CAC is comprised of nine members appointed by the state governor. However, some participants did not feel that the CAC was

[3] March 25, 2015, conference call with Doug Hindman, Kentucky CAC chair, and Craig Williams, Kentucky CDCAB co-chair; Judith Bradbury and Seth Tuler, committee members; Todd Kimmell, committee chair; and Jim Myska, study director.

"diverse enough to represent a broader swath of the community." An additional group, the CDCAB, a subcommittee of the CAC, was established. The CAC chair and the CDCAB co-chair are former members of the ACWA Dialogue group. According to the CDCAB co-chair, it

> includes representation of a more diverse group of people, so we could touch almost every aspect of life in the region. Now, the advisory board includes CAC members appointed by the governor and about 15 other people that represent different entities in the region not appointed by the governor—emergency preparedness, civic reps from other communities, hospitals, universities, chambers of commerce, NAACP. So, at meetings there is someone there that represents all aspects of life . . . religion, business, etc. so we have a link to all communities.[4]

The CAC also has working groups that focus on specific issues, such as the Secondary Waste Working Group. In addition, the CDCAB co-chair has been appointed as Madison County Chemical Weapons Host Liaison.

The goals and desired outcomes of BGCAPP's robust public involvement activities are as follows:

> To develop and maintain relationships with the local community, including CAC/CDCAB members, elected officials, oversight/regulatory stakeholders and the internal workforce to cultivate trust and understanding." The desired outcome is to develop an informed and educated stakeholder community that actively supports the program mission, [and]
>
> To provide consistent opportunities for public involvement and encourage community participation with the project. [The desired outcome is to develop] active partnerships within the stakeholder community to promote understanding of program decisions and requirements for success.[5]

Public involvement experiences at BGCAPP have resulted in strong local community ownership of the technological approach to destroying chemical agents at the BGCAPP, coupled with continuing opposition to offsite transport and disposal of wastes. These preferences have been articulated in the CAC/CDCAB meetings and public documents.

CONTINUED COLLABORATION AND TRUST IS CONTINGENT

"Critical trust" is an apt description of recent relationships among PEO ACWA, BGCAPP, and community representatives on the CAC and CDCAB. Poortinga and Pidgeon (2003, p. 971) call critical trust "a practical form of reliance on a person or institution combined with some healthy skepticism." It can serve important social functions, such as ensuring oversight and vigilance (see also Tuler, 2002). The CAC chair has said "We have a foundation of trust now." In parallel to the development of trust between the CAC/CDCAB, PEO ACWA, and BGCAPP, the general public in the community appears to trust the CAC/CDCAB. However, in this case trust lacks the critical dimension described by Poortinga and Pidgeon (2003). As described by both the CDCAB co-chair and the local BGCAPP outreach staff, "people feel the chairs represent them, and so they don't need to go to meetings. They trust these people who represent them."

The goals of BGCAPP's public involvement activities centered around the CAC/CDCAB are responsive "directly to citizens' calls for a decision process that acknowledges and integrates their ideas and concerns in the pursuit of technically sound and publicly acceptable approaches for weapons disposal" (Futrell, 2003a, p. 465). For example, while describing how the Army and CAC/CDCAB interact more recently, the CDCAB co-chair explained:

> We discussed how to estimate conditions of the mustard rounds here, and based on that, determine what would be [the] best path forward. The Army decided to do x-rays, determined that 60 percent or so would not drain. That would inhibit the throughput of the plant and impact downstream processing. What should we do? [An explosive destruction technology] was one option, and the [NRC] committee on this briefed us. We did [our own] research, etc. The CAC, over 2.5 years, discussed and learned, and then we agreed—because risks to workers outweighed the risks associated with deploying an explosive technology. Then we discussed the EDTs on the market. The CAC assessed different options. The Army did too, and then we discussed and agreed. It is a back and forth relationship—if there is an issue that needs to be raised by us or them, we work through it incrementally and work toward reaching agreement about how to proceed."[6]

At the same time, the foundation of trust is contingent on the character of ongoing interactions. Decisions and actions that appear to community members as violations of commitments or surprises have resulted in tensions in the consultative process and explicit opposition by community groups and representatives. For example, offsite shipments are viewed as ethically problematic, and failure to address the ethical dimensions of decisions to ship wastes offsite have created tensions in relationships that have improved since, in the words of the CAC chair, "the bad old days." For example, the CWWG actively opposed shipments of

---

[4] March 25, 2015, conference call with Doug Hindman, Kentucky CAC chair, and Craig Williams, Kentucky CDCAB co-chair; Judith Bradbury and Seth Tuler, committee members; Todd Kimmell, committee chair; and Jim Myska, study director.

[5] M. Monteverde, PEO ACWA, Public Affairs Office, and S. Parrett, public affairs officer, PEO ACWA, Blue Grass Chemical Activity, "Stakeholder Interests and Views on Offsite Shipment of Hydrolysate: Key Issues at the Site in Past," presentation to the committee on January 28, 2015.

[6] March 25, 2015, conference call with Doug Hindman, Kentucky CAC chair, and Craig Williams, Kentucky CDCAB co-chair; Judith Bradbury and Seth Tuler, committee members; Todd Kimmell, committee chair; and Jim Myska, study director.

hydrolysates from the Newport Chemical Agent Disposal Facility in Indiana because of concerns about environmental justice. In this case, potential recipient communities in Ohio and Delaware were opposed to receiving and processing the wastes, and they were supported in their opposition by the CWWG and others. The CWWG filed legal suits to stop the shipments and organized opposition. Ultimately, the hydrolysate was sent to Port Arthur, Texas, for incineration, although "people there were also opposed. But they did not have political clout, it was an environmental justice community, and they got it."[7]

In contrast to the experience with the shipments from the Newport Chemical Agent Disposal Facility, stakeholders were able to work out an acceptable arrangement for removing deteriorating steel containers storing a mixture of GB (sarin) nerve agent and its breakdown products from the BGAD.[8] The material was shipped to Port Arthur, Texas, in 2008 and 2009. The CAC/CDCAB agreed to these shipments as a "one time solution" that should not be viewed as a precedent for future shipments. In 2008 the CAC/CDCAB stated as follows:

> Notwithstanding our long opposition to offsite shipment, but recognizing the urgent risks of continued storage of the [ton containers], we are willing to tolerate offsite disposal of secondary wastes if the responsible agency (PEO ACWA) determines that on-site considerations would, for technical, regulatory, safety or other reasons, inhibit the expeditious elimination of the urgent risks associated with these materials. . . . Tolerating this one-time, offsite shipment of material the CAC/CDCAB does not in any way imply support for, the condoning of, or even consideration of any future similar shipments of similar materials offsite associated with the Blue Grass Chemical Agent Pilot Plant (BGCAPP).[9]

Opposition to offsite shipments was also expressed by local officials. In 2008 the chief executive of Madison County and the mayors of the cities of Richmond and Berea, Kentucky, issued a statement regarding offsite hydrolysate treatment:

> It is with steadfast resolve and unshakable determination, that the County and City Governments of Madison County and the Cities of Richmond and Berea join the Colorado and Kentucky Citizens Advisory Commissions and the Pueblo, Colorado Board of County Commissioners in opposing the offsite shipment of agent hydrolysate from either of the two ACWA chemical disposal facilities.[10]

According to the CDCAB co-chair, agreement about the Operation Swift Solution offsite shipments was achieved because of the meaningful participation of the CAC/CDCAB in discussions about the need for these shipments and planning for them.[11] The Kentucky CAC/CDCAB was provided the opportunity to understand the rationale for the decision, including community safety. It was also important that the recipient community was engaged in the planning process. The CAC/CDCAB described what happened this way:

> The community was engaged with our assistance, and we worked out an acceptable approach to them [Port Arthur, Texas, community and facility] receiving the material, with no protests, no lawsuits, no opposition, no politics. It worked because it was a smaller amount but more importantly they were part of a process before the decision was made.[12]

Yet, tensions remain because the possibility of shipping of hydrolysates offsite has been repeatedly raised by PEO ACWA during the past decade. This has been viewed with suspicion among some interested and affected parties because they understand that a commitment had been made to treat all wastes in a closed cycle onsite (NRC, 2015). The reasons given by the Army are not necessarily about protecting public or worker safety, but about cost savings, and the cost factor has been emphasized by Congress (in PL 110-417 National Defense Authorization Act for FY 2009), by the National Research Council, and by consultants (Noblis, 2008).

SUMMARY

Consistent with their descriptions of the consultation process for Operation Swift Solution, recent descriptions by the CAC chair and the CDCAB co-chair of the public involvement at BGCAPP emphasize their views that PEO ACWA is continuing to meet the community's expectations for meaningful public involvement. The process continues to unfold, but it is now based on a foundation of trust, commitment to transparency, and discussion of issues. For example, while acknowledging the initial lack of trust and the antagonistic relationship between the community and the Army, the CAC chair stated that the establishment of ACWA and its commitment to meeting public involvement goals allowed trust "to

---

[7] March 25, 2015, conference call with Doug Hindman, Kentucky CAC chair, and Craig Williams, Kentucky CDCAB co-chair; Judith Bradbury and Seth Tuler, committee members; Todd Kimmell, committee chair; and Jim Myska, study director.

[8] Operation Swift Solution, http://www.peoacwa.army.mil/bgcapp/about-bgcapp/operation-swift-solution/.

[9] M. Monteverde, PEO ACWA, Public Affairs Office, and S. Parrett, public affairs officer, PEO ACWA, Blue Grass Chemical Activity, "Stakeholder Interests and Views on Offsite Shipment of Hydrolysate: Key Issues at the Site in Past," presentation to the committee on January 28, 2015.

[10] Ibid.

[11] Operation Swift Solution involved the disposal of three deteriorating ton containers that held GB agent and related breakdown product. The GB in these containers was chemically neutralized in a special facility at BGAD and the hydrolysate was shipped offsite for final disposal, http://www.peoacwa.army.mil/bgcapp/about-bgcapp/operation-swift-solution/.

[12] March 25, 2015, conference call with Doug Hindman, Kentucky CAC chair, and Craig Williams, Kentucky CDCAB co-chair; Judith Bradbury and Seth Tuler, committee members; Todd Kimmell, committee chair; and Jim Myska, study director.

build up over time. The local BGCAPP folk have continued that tradition, helping to make things move forward. They are trying very hard to be transparent with us, we work hard to be transparent with them."[13]

Furthermore, the CAC chair and CDCAB co-chair reiterated that "offsite shipment is not supported, but we are forced to deal with the chance that things won't work, so it's important to have a backup plan, just in case."[14] They also outlined very explicitly their expectation for a similar level of involvement as with the Operation Swift Solution shipments and the current back-and-forth mode of decision making at BGCAPP:

> This is something we want to work out with them. We want to be part of that process. . . . We want to work it out with them. There are so many details involved—as your committee knows—there are so many pieces to this that it is impossible to answer [whether and how shipments should occur] now. But we are confident that with reason and an adequate level of data and information about why we think this and why we think that, ultimately we will reach a conclusion that is palatable to all stakeholders. . . . We need to be conscientious about all aspects of what happens if we are going to execute Plan B—we need to understand all dimensions and all impacts of what we are deciding.[15]

As the CAC chair stated in referring to the ACWA Dialogue experience, the key to resolving the issues posed as operations proceed is this:

> ACWA's transparency and willingness to work with the public was a major factor in managing opposition to incineration and involving the public in alternate technologies. I appreciated the opportunity to have input and to have my questions answered openly, which helped me to feel much more comfortable with the program's direction. (BGCAPP, 2015)

REFERENCES

CWWG (Chemical Weapons Working Group). 2007. *Citizen Groups File Summary Judgment Motion in Federal Court to Stop VX Waste Shipments to Texas Incinerator*. http://cwwg.org/pr_10.23.07sjmotion.html.

Durant, R.F. 2007. *The Greening of the U.S. Military*. Washington, D.C.: Georgetown University Press.

Futrell, R. 2003a. Technical adversarialism and participatory collaboration in the U.S. chemical weapons disposal program. *Science, Technology, & Human Values* 28(4): 451-482.

Futrell, R. 2003b. Framing processes, cognitive liberation, and NIMBY, Protest in the U.S. chemical-weapons disposal conflict. *Sociological Inquiry* 73(3): 359-386.

Futrell, R., and Futrell, D. 2012. Expertise and alliances: How Kentuckians transformed the U.S. chemical wapons disposal program. S. McSpirit, L. Faltraco, and C. Bailey (eds.), *Confronting Ecological Crisis in Appalachia and the South: University and Community Partnerships*, pp. 171-193. Lexington, Ky.: University Press of Kentucky.

GAO (General Accounting Office). 1995a. *Chemical Stockpile Disposal Program Review*. NSIAD-95-66R. Washington, D.C.: General Accounting Office.

GAO. 1995b. *Chemical Weapons Disposal: Issues Related to DOD's Management*. T/NSIAD-95-18. Washington, D.C.: General Accounting Office.

Goldberg, M. 2003. Strengthening the link between project planning and environmental impact assessment: The assembled chemical weapons assessment dialogue process. *Environmental Practice* 5(4): 313-320.

Keystone Policy Center. 2004. *ACWA Dialogue Close-out Report* (draft) https://www.keystone.org/.

Marshall, S. 1996. *Chemical Weapons Disposal and Environmental Justice*. Report by the Kentucky Environmental Foundation.

Noblis. 2008. *Offsite Disposal of ACWA Hydrolysates*. NTR 2008-61129. Falls Church, Va.

NRC (National Research Council). 2015. *Review Criteria for Successful Treatment of Hydrolysate at the Pueblo Chemical Agent Destruction Pilot Plant*. Washington, D.C.: The National Academies Press.

PEO ACWA (Program Executive Office, Assembled Chemical Weapons). *Program Executive Office, Assembled Chemical Weapons,* Facts page. http://www.peoacwa.army.mil/media-toolkit/facts-pages/peo-acwa/.

Poortinga, W., and N.F. Pidgeon. 2003. Exploring the dimensionality of trust in risk regulation. *Risk Analysis* 23(5): 961-972.

Tuler, S. 2002. *Radiation Risk Perception and Communication: A Case Study of the Fernald Environmental Management Project*. SERI Report 02-005. Greenfield, Mass.: Social and Environmental Research Institute. http://www.seri-us.org/sites/default/files/fernald_0.pdf.

[13] Ibid.
[14] Ibid.
[15] Ibid.

B

Public Interest and Input Documents

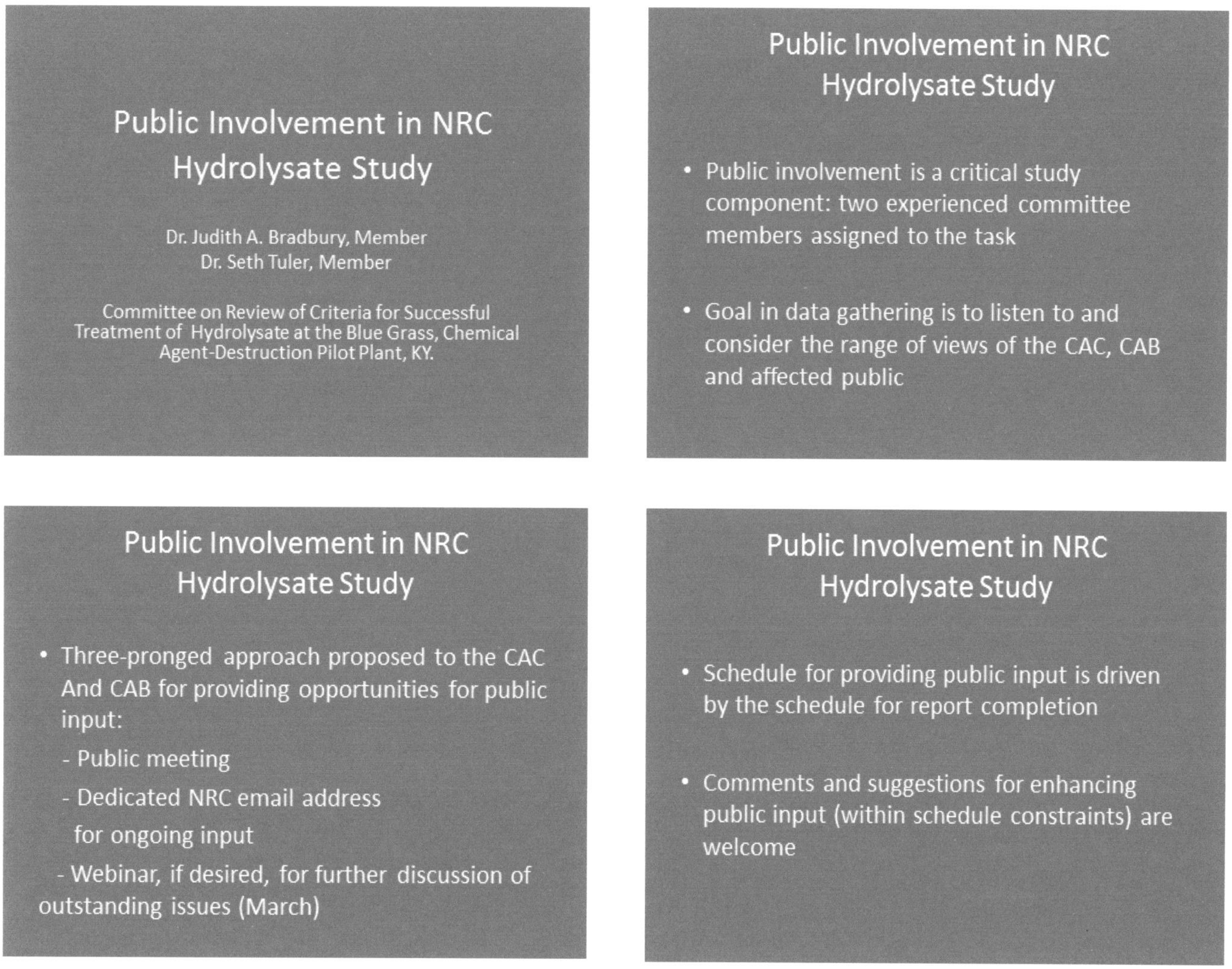

Exhibit 1. Committee approach to encouraging public input. Presented at committee meeting on January 29, 2015.
(Exhibit continues on next page.)

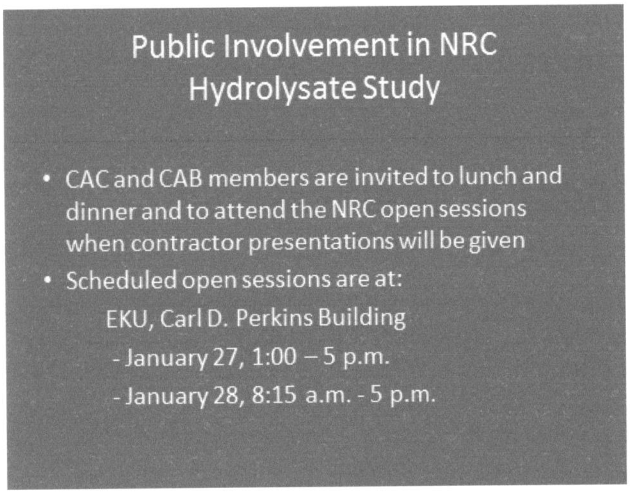

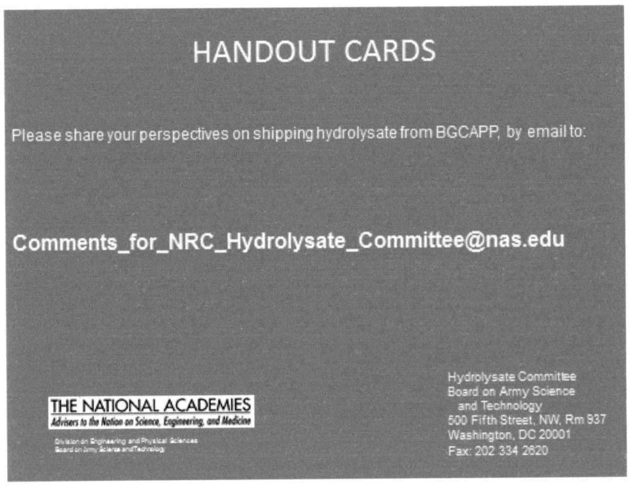

Exhibit 1. Continued.

1/21/2015 NRC seeks public comment on hydrolysate transport Richmond Register: News

http://www.richmondregister.com/news/nrcseekspubliccommentonhydrolysatetransport/article_ead95fb8a1cd11e4a983bf9e2a947b10.html?mode=print 1/1

NRC seeks public comment on hydrolysate transport

Special to the Register | Posted: Wednesday, January 21, 2015 7:30 pm
A National Research Council (NRC) committee is seeking public comment regarding the potential for the shipment of hydrolysate from the Blue Grass Chemical Agent Destruction Pilot Plant after chemical weapons agents are destroyed.

Datagathering sessions will be conducted at the Eastern Kentucky University Perkings Building from 1 to 5 p.m. Tuesday and Wednesday from 8 a.m. to 5 p.m. A public comment session is scheduled for 7 p.m. Wednesday.

During the two daytime sessions, the committee will hear presentations from and hold discussions with government BGCAPP personnel; the contractor, Bechtel Parsons; personnel from the Kentucky Department for Environmental Protection, and representatives of the Citizens' Advisory Commission and the Community Advisory Board.

In addition, the committee has established a dedicated email address for receiving public comments: Comments_for_NRC_Hydrolysate_Committee@nas.edu.

The plant is designed to destroy the stockpile of chemical weapons stored at the Army's Blue Grass Army Depot. Hydrolysate is the waste product that remains following the chemical destruction by hydrolysis of the chemical agents and energetics.

The Program Manager for Assembled Chemical Weapons Alternatives has tasked the NRC committee to independently develop criteria for safely and successfully treating the hydrolysate that will be produced at BGCAPP, and, if the supercritical water oxidation system underperforms in treating hydrolysate, to identify potential modifications that would allow continued onsite processing, as well as regulatory requirements for offsite hydrolysate and treatment options.

"Part of our tasking is to consider stakeholder interests and solicit stakeholder input," said committee chair Todd A. Kimmell, principal investigator with the Environmental Science Division at Argonne National Laboratory. "We want to understand the perspectives of the people who live in the surrounding area and who work at BGCAPP."

Exhibit 2. Article from *Richmond Register* website, January 21, 2015, announcing committee meeting, public meeting, and e-mail address for submitting public comments to the committee.

From: Doug Hindman
Sent: Tuesday, February 03, 2015 5:37 PM
To: Hydrolysate Committee
Subject: Hydrolysate comment

As chair of the KY CAC I was privileged to address your committee and to attend at least part of your Kentucky meeting.

I'm concerned that your committee is unduly restricting the options you are considering. Your statement of task asks you "to identify potential modifications that would allow continued onsite processing." The only potential modification I heard addressed was off-site transportation. In fact, your card seeking public comment asks for perspectives only on this option. While it's an obvious option, I hope your committee will also consider other, especially on-site, options.

For example, I asked you to consider use of the EDT presently under construction at Kentucky's pilot plant. This EDT will be a one-step treatment for mustard-filled projectiles that should finish before the main plant begins. Thus it could be available to treat "excess" hydrolysate if the SCWO system develops problems.

I do not expect your committee to evaluate such options in detail. I urge you to identify possible on-site options, such as EDT, for more detailed evaluation by ACWA.

Thanks to your committee for your efforts.

Douglas Hindman

Exhibit 3. E-mail from Douglas Hindman, Chair, Kentucky Citizens' Advisory Commission, on February 3, 2015.

Kentucky Chemical Demilitarization Citizens' Advisory Commission

Chemical Demilitarization Citizens' Advisory Commission
Chemical Destruction Community Advisory Board
105 5th Street, Suite 206
Richmond, KY 40475
859.624.4700 / 859.986.7565

CDCAB
CHEMICAL DESTRUCTION COMMUNITY ADVISORY BOARD

Doug Hindman
Chair

Reagan Taylor
Craig Williams
Co-Chairs

NRC Hydrolysate Committee
500 Fifth Street, NW, Rm 937
Washington, D.C. 20001

March 2, 2015

Dear Committee Members,

Herein please find the salient points raised during your Committee meeting held in Richmond, Kentucky in January of this year.

Again, please accept our appreciation for holding the meeting here in the affected community.

<u>OVERARCHING POSITIONS</u>

1) This study is being undertaken strictly as a contingency plan should hydrolysate secondary waste processing problems necessitate off-site shipment; and

2) Every opportunity for CAC/CDCAB involvement and input should be provided and welcome.

<u>SUGGESTED CRITERIA</u>

<u>IF</u> secondary treatment of hydrolysate is unable to keep up with neutralization output (due to unexpected processing limitations with the SCWO system, etc.);

<u>AND</u> if all feasible mitigation measures have been evaluated;

<u>AND</u> if mitigation is impossible or would require that plant operations be suspended for an extensive period of time. ("Extensive" should be defined within the Study)

<u>THEN</u> alternative approaches ("Plan B") would be acceptable <u>BUT ONLY</u> for that portion of the neutralization output that cannot be processed through the planned secondary treatment and/or stored on site until mitigation is achieved. It's easy to conceive that problems may be temporary or limited so that only part of the hydrolysate would need transportation).

<u>ALSO</u>, The NRC should consider alternative on-site treatment possibilities. For example, it might be possible to modify the EDT to process "excess" hydrolysate.

<u>SUGGESTED STUDY ELEMENTS</u>

It is our expectation that the NRC will:

Exhibit 4. Statement from the Kentucky Citizens' Advisory Commission and Chemical Destruction Community Advisory Board, March 2, 2015.
(Exhibit continues on next page.)

- Highlight at the BEGINNING of the report that the effort is being made only as a contingency in the event that the on-site treatment reaches a point where there is NO OTHER OPTION than to ship off-site.

- Work with ACWA, BPBG, KDEP, CAC/CDCAB, and BGAD to develop a set of protocols that will identify under what circumstances off-site shipment would occur. These should include the study objectives in the Statement of Task and also include, but not be limited to:

• Consideration of the historical position of the CAC/CDCAB and reception communities of past agent hydrolysate shipments;

• Identification of technology and location of any potential reception site/community;

• Consideration of schedule impact via opposition (regulatory, legal, political) from identified potential reception site(s);

• Developing a means of information sharing and input from the CAC/CDCAB as items identified in the Statement of Task are evolving. (i.e. storage levels identified that might be a threshold to trigger shipment);

• Providing certain CAC/CDCAB members review in draft form opportunities in accordance with procedures approved by the NRC Report Review Committee, similar to the Citizens Advisory Technical Team (CATT) associated with the ACWA Dialogue.

Sincerely,

Doug Hindman *Craig Williams*

Doug Hindman
Chair, KY CAC

Craig Williams
Co-Chair, KY CDCAB

Exhibit 4. Continued.

C

Biographical Sketches of Committee Members

Todd A. Kimmell, *chair*, is principal investigator with the Environmental Science Division at the U.S. Department of Energy's Argonne National Laboratory. He is an environmental scientist and policy analyst with more than 30 years' experience in solid and hazardous waste management, permitting and regulatory compliance, cleanup programs, environmental programs policy development, and emergency management and homeland security. He has supported the Army's chemical and conventional munitions management programs and has contributed to the Army's Assembled Chemical Weapons Assessment program and the Chemical Stockpile Emergency Preparedness Program. Mr. Kimmell also has a strong technical background in analytical and physical/chemical test method development and analytical quality assurance and control. Mr. Kimmell has also supported a number of environmental permitting programs at Army chemical weapons storage sites and at open burning/open detonation sites. He graduated from George Washington University with an M.S. in environmental science.

Edward A. Bouwer is currently the Abel Wolman Professor of Environmental Engineering and chair of the department of Geography and Environmental Engineering at Johns Hopkins University. He is also director of the Center for Contaminant Transport, Fate and Remediation. Prior to this position, Dr. Bouwer spent 7 years as director of the Center for Hazardous Substances in Urban Environments, a project that was funded by the U.S. Environmental Protection Agency. Dr. Bouwer's research interests encompass factors that influence biotransformation of contaminants, bioremediation for control of contaminated soils and groundwaters, biofilm kinetics, biological processes design in wastewater, industrial, and drinking water treatment, transport and fate of microorganisms in porous media, and the behavior of metal and organic contaminants in sediments and aquatic ecosystems. Dr. Bouwer received his B.S.C.E. in civil engineering with a minor in nuclear engineering from Arizona State University and an M.S. and a Ph.D. in environmental engineering and science from Stanford University

Judith A. Bradbury graduated from the University of Pittsburgh with a Ph.D. in public and international affairs and has an M.A. in public affairs from the Indiana University of Pennsylvania and a B.S. in sociology from the London School of Economics. She retired after almost 20 years as a senior social scientist with the Pacific Northwest National Laboratory, which is operated by Battelle for the U.S. Department of Energy. In her work, she has emphasized the relevance of social science insights and tools to the analysis and resolution of science policy issues. She has extensive experience in both the practice of and research into public involvement and institutional activities. Her experience includes (most recently) responsibility for planning and implementing outreach and education activities for the Midwest Regional Carbon Sequestration partnership. Previous work includes evaluation of selected U.S. Army Restoration Advisory Boards; a series of evaluations of the effectiveness of Department of Energy's (DOE's) Site-Specific Advisory boards; evaluation of training programs in public participation for DOE managers; meeting facilitation, planning, and program evaluation for the DOE nuclear waste transportation program; and research into community perspectives on the risk of incineration for disposing of the nation's stockpile of chemical weapons.

Rebecca A. Haffenden, Esq., is an attorney and currently serves as a program's attorney at the Argonne National Laboratory. Her recent professional work has included work for the U.S. Department of Homeland Security to evaluate legislation and regulations associated with security vulnerabilities and providing legal expertise to programs involving federal facility site remediation and hazardous waste compliance and corrective actions (Resource Conservation and Recovery Act). She also coauthored a working paper on the application of federal and state hazardous waste regulatory programs to waste chemical agents, in addition to being a co-author of the Environmental Impact Statement for the Assembled Chemical Weapons Alternatives program. Ms. Haffenden received a B.A. in psychology from the University of Illinois and a J.D. degree from Suffolk Law School, Boston, Massachusetts.

Kimberly L. Jones is a professor and chair of the Department of Civil Engineering at Howard University. She previously worked as an associate and assistant professor in this department from 1996 to 2009. Over the past 5 years, her research objectives have primarily been interdisciplinary, collaborative research in the emerging research areas of nanotechnology and nanobiotechnology, while continuing to build her environmental engineering capabilities. She has worked to develop an effective research strategy to investigate innovative technologies involving nanotechnology, environmental engineering, and membrane processes in an effort to solve some of the more pervasive problems facing our world, while working to attract, retain, and graduate technically competent African-American students to increase the number of minority engineers and scientists in academic-, industrial-, and government-related careers. Dr. Jones received her B.S. in civil engineering from Howard University, an M.S. in civil and environmental engineering from the University of Illinois, and a Ph.D. in environmental engineering from the Johns Hopkins University.

Murray Glenn Lord is associate environmental health and safety (EH&S) director in the EH&S Operations Technology Center at Dow Chemical Company. He is responsible for the research program for technology development for Global Environmental Operations, which includes project areas in process optimization, technology development, and capital project execution. Mr. Lord has experience in project areas across multiple business and technology areas. He is also accountable for EH&S performance, budget performance, project development, and personnel leadership of a research group from four locations and is the leader of the Environmental Technology Leadership Group, accountable for environmental technology development for Dow. Previously, Mr. Lord was a technical leader of propylene oxide process research and was responsible for a research program in support of technology development of the propylene oxide process. He was also responsible for development and coordination of research studies at laboratory, pilot plant, and full commercial scale.

Douglas M. Medville retired from The MITRE Corporation as program leader for chemical materiel disposal and remediation. He has led many analyses of risk, process engineering, transportation, and alternative disposal technologies and has briefed the public and senior military officials on the results. Mr. Medville was responsible for evaluating the reliability and performance of the demilitarization machines used by the Army to disassemble stockpile chemical munitions and wrote several test plans and protocols for alternative chemical munition disposal technologies. He also led the evaluation of the operational performance of the Army's chemical weapon disposal facility on Johnson Atoll and directed an assessment of the risks, public perceptions, environmental aspects, and logistics of transporting recovered nonstockpile chemical warfare materiel to candidate storage and disposal destinations. Following his retirement from MITRE, he participated as a committee member in 10 NRC studies concerning the Army's ACWA and nonstockpile programs and was vice chair for three of these committees. Prior to his work at MITRE, Mr. Medville worked as an engineer for the Franklin Institute Research Laboratories and General Electric. Mr. Medville earned a B.S. in industrial engineering and an M.S. in operations research, both from New York University.

Trisha H. Miller is a systems analyst/engineer with Sandia National Laboratories. She has participated in a number of analysis projects focused on chemical security, including projects supporting the Department of Homeland Security related to the evaluation of the security benefits of inherently safer technologies in the chemical industry and risk assessments for chemical attacks. Dr. Miller was awarded an early career grant to develop new methodologies for end-to-end analysis of chemical defense systems. She received her Ph.D. in chemistry from the University of California at Berkeley in 2009. She serves as an adjunct faculty member in the Department of Chemistry at Augsburg College in Minneapolis, Minnesota.

Robert B. Puyear is an independent consultant specializing in corrosion prevention and control, failure analysis, and materials selection. Mr. Puyear worked at the Haynes Stellite Division of Union Carbide for 16 years developing high-performance materials for chemical and aerospace applications. He also worked for Monsanto for 21 years as a corrosion specialist, where he managed the mechanical and materials engineering section. He is an expert in materials engineering and evaluating materials of construction. Mr. Puyear graduated from the Missouri School of Mines and Metallurgy with a B.S. in chemical engineering and from Purdue University with an M.S. in industrial administration. He was also a member of the National Research Council Committee on Review and Evaluation of the Army Chemical Stockpile Disposal Program.

William R. Rhyne is a retired risk and safety analysis consultant to the nuclear, chemical, and transportation industries, He has over 30 years' experience associated with nuclear and chemical processing facilities and with the transportation of hazardous materials. From 1984 to 1987, he was the project manager and principal investigator for a probabilistic analysis of transporting obsolete chemical munitions. Beginning in 1997, he was a member of several NRC committees for the Assembled Chemical Weapons Alternatives program and is a former member of the Committee on Chemical Demilitarization (2007-2010). Dr. Rhyne has authored or coauthored numerous publications and reports on nuclear and chemical safety and risk analysis areas and is the author of the book *Hazardous Materials Transportation Risk Analysis: Quantitative Approaches for Truck and Train.* He is a former member of the NRC Transportation Research Board's Hazardous Materials Committee, the Society for Risk Analysis, the American Nuclear Society, and the American Institute of Chemical Engineers. He received a B.S. in nuclear engineering from the University of Tennessee and M.S. and D.Sc. degrees in nuclear engineering from the University of Virginia.

Phillip E. Savage is the head of the chemical engineering department at Penn State. He earned a B.S. from Penn State and M.Ch.E. and Ph.D. degrees from the University of Delaware. All of his degrees are in chemical engineering. His research and teaching focus on the rates, mechanisms, and engineering of chemical reactions that move us toward a more environmentally sustainable society. Current research projects deal with hydrothermal reactions that can be used for hydrogen production from biomass and for liquid transportation fuel production from algae. His teaching focuses on chemical reaction engineering and environmental sustainability. Dr. Savage is editor in chief for *Industrial & Engineering Chemistry Research,* and he is on the editorial boards for the *Journal of Supercritical Fluids, Energy & Fuels,* and *Environmental Progress & Sustainable Energy.* Dr. Savage is a fellow of the American Institute of Chemical Engineers and the American Chemical Society. He received the 2009 Michigan Governor's Award for Green Chemistry and the 2001 National Catalyst Award from the American Chemistry Council in recognition of his outstanding teaching and contributions to chemical education.

Philip C. Singer (NAE) is an emeritus professor in the Department of Environmental Sciences and Engineering in the Gillings School of Global Public Health at the University of North Carolina at Chapel Hill (UNC), where he was the Dan Okun Distinguished Professor of Environmental Engineering from 2002 to 2010. After obtaining his Ph.D. from Harvard University in 1969, Dr. Singer was an assistant professor in the Department of Civil Engineering at the University of Notre Dame before joining the faculty at UNC in 1973. He conducted research on the chemical aspects of water and wastewater treatment and on aquatic chemistry for 45 years and has published more than 250 papers and reports on these subjects. Dr. Singer has been active in the American Water Works Association and has served on the National Research Council's Water Science and Technology Board and its Committee on Drinking Water Contaminants and on the U.S. Environmental Protection Agency's Science Advisory Board and the National Drinking Water Advisory Council. Dr. Singer is a recipient of the American Water Works Association's A.P. Black Research Award and the Abel Wolman Award of Excellence, the American Academy of Environmental Engineers' Gordon Maskew Fair Award, the National Water Research Institute's Athalie Richardson Irvine Clarke Prize, and the Association of Environmental Engineering and Science Professors' Charles R. O'Melia Distinguished Educator Award. He was elected to membership in the National Academy of Engineering in 2005. Dr. Singer is currently a part-time consultant with CDM-Smith.

Seth Tuler is an associate teaching professor in the Interdisciplinary and Global Studies Division, Worcester Polytechnic Institute. His research interests have focused on public participation, risk communication, risk governance, and developing tools to characterize human impacts and vulnerabilities to risk events. He seeks to apply insights emerging from research to practical applications in a wide range of policy arenas, including climate change adaptation planning, nuclear waste management, marine fisheries management, and cleanup of contaminated sites. He previously served on the National Academy of Science's Committee on Transportation of Spent Nuclear Fuel and High Level Radioactive Waste and the Federal Advisory Committee on Energy-Related Epidemiologic Research, chairing its Subcommittee for Community Affairs for 2 years, and an ad hoc committee to advise the National Cancer Institute in its efforts to inform people about health risks from iodine-131 nuclear weapons testing fallout. Dr. Tuler has an extensive publication record in peer-reviewed journals, book chapters, and peer-reviewed technical reports. He was a coauthor of two technical reports for President Obama's Blue Ribbon Commission on America's Nuclear Future. Dr. Tuler received a B.A. in mathematics from the University of Chicago, an M.S. in technology and policy from the interdisciplinary Technology and Policy Program of the Massachusetts Institute of Technology, and a Ph.D. from the Environmental Science and Policy Program, Clark University, Worcester, Massachusetts.

D

Committee Activities

This committee authored two reports, one on the Pueblo Chemical Agent Destruction Pilot Plant (PCAPP) and one on the Blue Grass Chemical Agent Destruction Pilot Plant (BGCAPP). The committee held three meetings while drafting the PCAPP report. Since those meetings have no bearing on the drafting of this BGCAPP report, the PCAPP-related meetings and activities are not listed in this appendix. They are listed in Appendix D of the PCAPP report, *Review Criteria for Successful Treatment of Hydrolysate at the Pueblo Chemical Agent Destruction Pilot Plant*. This listing includes only meetings and activities related to the development of this BGCAPP report, and the meeting count begins with the fourth meeting.

FOURTH MEETING JANUARY 27-29, 2015

Carl D. Perkins Building Eastern Kentucky University, Richmond, Kentucky, and Hyatt Regency Lexington, Lexington, Kentucky

Objectives: Conduct composition/balance/bias discussions for new committee members; discuss the statement of task and project background and review it with the sponsor; receive BGCAPP overview and process briefings; meet with regulators and public stakeholders; review report writing process and the project plan; refine the draft outline for the report; proceed toward the concept draft; make committee writing assignments; and decide on future meeting dates and next steps.

Safety Briefing and Site Tour of the Blue Grass Chemical Agent Destruction Pilot Plant

Discussion with Sponsor, Jeffrey Brubaker, BGCAPP site project manager

BGCAPP Project Overview, Rick Goetz, Parsons, BGCAPP assistant project manager

BGCAPP Agent Treatment Process, John Barton, Battelle, BGCAPP chief scientist

RCRA Permit Structure and Potential Modifications for Offsite Shipment of Hydrolysate, John McArthur, Parsons, BGCAPP environmental manager

Other Regulatory Requirements and Notifications Applicable to Offsite Shipment of Hydrolysate, John McArthur, Parsons, BGCAPP environmental manager

BGAD RCRA Permit from the Perspective of the Kentucky Department for Environmental Protection (KDEP): Regulatory and Other Requirements and Notifications Applicable to Offsite Shipment, Tim Hubbard, assistant director, Division of Waste Management, Kentucky Department for Environmental Protection

Committee Approach to Soliciting Stakeholder Input, Judith Bradbury, member, Hydrolysate Committee

Stakeholder Interests and Views on Offsite Shipment of Hydrolysate: Key Issues at the Site in Past, Miguel Monteverde, PEO ACWA, public affairs office, and Stephanie Parrett, PEO ACWA, Blue Grass Chemical Activity, public affairs officer

Stakeholder Interests and Views on Offsite Shipment of Hydrolysate, Craig Williams, co-chair, Chemical Destruction Community Advisory Board, and Douglas Hindman, chair, Citizens' Advisory Council

Supercritical Water Oxidation Process: Cradle to Grave, George Lucier, Battelle, BGCAPP deputy chief scientist, and Larry Austin, Parsons, BGCAPP waste manager

Supercritical Water Oxidation Risk Mitigation Activities, George Lucier, Battelle, BGCAPP deputy chief scientist

Downstream Impacts to Plant Operations If Offsite Shipment Is Required, John Barton, Battelle, BGCAPP chief scientist

PUBLIC MEETING TO SOLICIT LOCAL STAKEHOLDER INPUT JANUARY 29, 2015

Carl D. Perkins Building Eastern Kentucky University, Richmond, Kentucky

Attended by the whole committee, led by Todd Kimmell, chair, and Judith Bradbury, member.

TELECONFERENCE WITH BGCAPP TO DISCUSS SCWO MARCH 13, 2015

NRC Participants

Edward Bouwer, member, Hydrolysate Committee
Rebecca Haffenden, member, Hydrolysate Committee
Kimberly Jones, member, Hydrolysate Committee
Todd Kimmell, chair, Hydrolysate Committee
Glenn Lord, member, Hydrolysate Committee
Douglas Medville, member, Hydrolysate Committee
Trisha Miller, member, Hydrolysate Committee
James Myska, program officer, National Research Council
Robert Puyear, member, Hydrolysate Committee
Phillip Savage, member, Hydrolysate Committee
Philip Singer, member, Hydrolysate Committee

BGCAPP Participants

John Barton, Battelle, BGCAPP chief scientist
Jeff Kresja, BGCAPP deputy site project manager, Compliance
George Lucier, Battelle, BGCAPP deputy chief scientist

FIFTH MEETING MARCH 18-20, 2015

Keck Center, Washington, D.C.

Objectives: Discuss data gathering, review and discuss the report draft, attain a first full-message draft, confirm committee writing assignments, and confirm the next steps in the report drafting process.

TELECONFERENCE WITH PUBLIC STAKEHOLDERS MARCH 26, 2015

NRC Participants

Judith Bradbury, member, Hydrolysate Committee
Todd Kimmell, chair, Hydrolysate Committee
James Myska, program officer, National Research Council
Seth Tuler, member, Hydrolysate Committee

Stakeholder Participants

Doug Hindman, chair, Kentucky Chemical Demilitarization Citizens' Advisory Commission
Craig Williams, co-chair, Kentucky Chemical Destruction Community Advisory Board

TELECONFERENCE WITH BGCAPP AND ACWA PUBLIC AFFAIRS APRIL 7, 2015

NRC Participants

Judith Bradbury, member, Hydrolysate Committee
Todd Kimmell, chair, Hydrolysate Committee
James Myska, program officer, National Research Council
Seth Tuler, member, Hydrolysate Committee

BGCAPP and ACWA Participants

Miguel Monteverde, ACWA public affairs officer
Sarah Parke, Blue Grass Chemical Activity public affairs officer

SIXTH MEETING MAY 19-20, 2015

Keck Center, Washington, D.C.

Objectives: Conduct thorough review of preconcurrence draft, accomplish any last writing needed, and achieve committee concurrence.